Lab Manual

TO ACCOMPANY

Practical Heating Technology

WILLIAM M. JOHNSON
Central Piedmont Community College
Charlotte, North Carolina

Delmar Publishers Inc.™

I(T)P™ An International Thomson Publishing Company

New York • London • Bonn • Boston • Detroit • Madrid • Melbourne • Mexico City • Paris
Singapore • Tokyo • Toronto • Washington • Albany NY • Belmont CA • Cincinnati OH

NOTICE TO THE READER

Publisher does not warrant or guarantee any of the products described herein or perform any independent analysis in connection with any of the product information contained herein. Publisher does not assume, and expressly disclaims, any obligation to obtain and include information other than that provided to it by the manufacturer.

The reader is expressly warned to consider and adopt all safety precautions that might be indicated by the activities described herein and to avoid all potential hazards. By following the instructions contained herein, the reader willingly assumes all risks in connection with such instructions.

The publisher makes no representations or warranties of any kind, including but not limited to, the warranties of fitness for particular purpose or merchantability, nor are any such representations implied with respect to the material set forth herein, and the publisher takes no responsibility with respect to such material. The publisher shall not be liable for any special, consequential or exemplary damages resulting, in whole or in part, from the readers' use of, or reliance upon, this material.

COVER PHOTOS: Courtesy of Lennox Industries Inc. and Weil-McLain

DELMAR STAFF
    Senior Administrative Editor: Vernon Anthony
    Project Editor: Eleanor Isenhart
    Production Coordinator: Dianne Jensis
    Art/Design Coordinator: Heather Brown

COPYRIGHT © 1994
By Delmar Publishers Inc.
an International Thomson Publishing Company

The ITP logo is a trademark under license

Printed in the United States of America

For more information, contact:

Delmar Publishers Inc.
3 Columbia Circle, Box 15015
Albany, New York 12212-5015

International Thomson Publishing
Berkshire House
168-173 High Holborn
London, WC1V7AA
England

Thomas Nelson Australia
102 Dodds Street
South Melbourne 3205
Victoria, Australia

Nelson Canada
1120 Birchmont Road
Scarborough, Ontario
M1K 5G4, Canada

International Thomson Publishing GmbH
Konigswinterer Str. 418
53227 Bonn
Germany

International Thomson Publishing Asia
221 Henderson Bldg. #05-10
Singapore 0315

International Thomson Publishing Japan
Kyowa Building, 3F
2-2-1 Hirakawa-cho
Chiyoda-ku, Tokyo 102
Japan

All rights reserved. No part of this work covered by the copyright hereon may be reproduced or used in any form or by any means—graphic, electronic, or mechanical, including photocopying, recording, taping, or information storage and retrieval systems—without the written permission of the publisher.

1 2 3 4 5 6 7 8 9 10 XXX 00 99 98 97 96 95 94

Library of Congress Catalog Card Number: 94-251
ISBN: 0-8273-4883-5

Contents

Preface iv

Tool List iv

Lab 1   Properties of Gases 1
Lab 2   Gas Burners and Heat Exchangers 3
Lab 3   Gas Furnace Familiarization 5
Lab 4   Identification of the Pilot Safety Feature of a Gas Furnace 7
Lab 5   Gas Furnace Heat Sensing Elements 9
Lab 6   Gas Pressure Testing 11
Lab 7   Orifice Sizing 13
Lab 8   Wiring for a Gas Furnace 15
Lab 9   Maintenance of Gas Burners 17
Lab 10 Checking a Gas Furnace for Correct Venting 19
Lab 11 Combustion Analysis of a Gas Furnace 21
Lab 12 Determining Airflow (CFM) Using the Air-Temperature-Rise Method 23
Lab 13 Changing a Gas Valve on a Gas Furnace 25
Lab 14 Changing a Fan Motor on a Gas Furnace 27
Lab 15 Properties of Oil 29
Lab 16 Familiarization with Oil Furnace Components 31
Lab 17 Oil Burner Maintenance 33
Lab 18 Electronic Controls for Oil Burners 35
Lab 19 Checking the Fan and Limit Control 37
Lab 20 Adjusting an Oil Burner Pump Pressure 39
Lab 21 Combustion Analysis of an Oil Burner 41
Lab 22 Oil Furnace, Changing the Nozzle 43
Lab 23 Oil Furnace, Changing an Oil Pump 45
Lab 24 Making Voltage and Amperage Readings with a VOM 47
Lab 25 Measuring Resistance and Using Ohm's Law 51
Lab 26 Checking the Accuracy or Calibration of Electrical Instruments 53
Lab 27 Electric Furnace Familiarization 57
Lab 28 Determining Airflow (CFM) by Using the Air Temperature Rise 59
Lab 29 Setting the Heat Anticipator for an Electric Furnace 61
Lab 30 Low-Voltage Control Circuits Used in Electric Heat 63
Lab 31 Checking a Package Sequencer 65
Lab 32 Checking Electric Heating Elements Using an Ammeter 67
Lab 33 Changing a Sequencer 69
Lab 34 Changing the Heating Element in an Electric Furnace 71
Lab 35 Familiarization of a Hot Water Boiler and Heating System 73
Lab 36 Checking Water Flow Using an Orifice Flow Checking Device 75
Lab 37 Checking the System Flow Rate Using the Pressure Drop Across the Pump 77
Lab 38 Cleaning a Strainer 79
Lab 39 Filling a Hot Water Heating System 81
Lab 40 Filling an Expansion Tank on a Hot-Water Heating System 83
Lab 41 Installing a Pump Coupling 85
Lab 42 Familiarization of a Steam Boiler and Heating System 87
Lab 43 Checking a Condensate Trap for Proper Operation 89
Lab 44 Performing Blowdown on a Steam Boiler 91

PREFACE

The lab exercises in this manual are intended to enable the student to perform work exercises that are similar to actual field experiences. Some of the exercises are to be written only. The purpose of having the student perform calculations and answer questions is to stimulate interest while reinforcing the subject matter. Actual service exercises are included, when possible, so the student actually performs the exercises on equipment.

An effort was made to make the exercises as generic as possible. Most school systems should have the equipment used in these labs. The class may use equipment that the school uses to heat the building for some projects. No exercise in this text should cause a piece of equipment to be put out of service for any longer than it takes to complete the exercise.

A recommended tool list is included for the purpose of providing the lab instructor with the tools necessary to complete these labs. This list can also be used to help in choosing the tools to be used for the course.

Particular emphasis should be placed on the "Safety Precautions" that appear throughout the text and lab manual. Students should be required to adhere to safe lab and shop practices for their own protection and to form good safety habits.

Answers to the questions in the text and lab manual can be found in the Instructor's Guide, which accompanies *Practical Heating Technology*. You may also refer to the Instructor's Guide for suggestions regarding organization of the material and general teaching suggestions.

TOOL LIST

This tool list includes the tools required to complete the lab exercises in this book as well as some basic supplies. Brand names are used in some cases; this is not meant to be an endorsement of these particular tools. These have been used successfully for these exercises in our program.

- 2 Glass stem thermometers with a range from about –30 to 250°F. A 12" thermometer will give the best readings.
- 2 Dial type thermometers with ranges from –20 to 250°F
- 1 4 lead electronic thermometer
- 1 Pair of side shield goggles
- 1 Calculator for simple calculations
- 1 Pair of light gloves, cotton or leather
- 1 Medium straight blade screwdriver
- 1 Small control screwdriver
- 1 Medium Phillips head screwdriver
- 1 Pair of needle nose pliers with insulated handles
- 1 Pair of slip joint pump pliers
- 1 Pair of electrician's side cutting pliers, 6"
- 1 Pair of electrician's diagonal cutting pliers
- 2 4" adjustable wrenches, Diamond with extended opening
- 2 6" adjustable wrenches, Diamond with extended opening
- 2 8" adjustable wrenches, Diamond with extended opening
- 2 10" pipe wrenches
- 2 12" pipe wrenches
- 1 Soft face hammer
- 1 Assortment of Allen head wrenches
- 1 Assortment of 3/8" drive socket wrenches
- 1 Assortment of open end wrenches, 3/8" through 3/4"
- 1 Assortment of box end wrenches, 3/8" through 3/4"
- 1 Inspection mirror
- 1 Quality flashlight, plastic preferred to prevent shock hazard
- 1 Assortment of nut drivers
- 1 Set of tubing tools, cutter-reamer and flaring tool
- 1 Set of wire brushes for cleaning tubing
- 1 Bench vise
- 1 Resistance heater, such as a space heater
- 1 Assortment of resistors, gold band preferred
- 1 Clamp on ammeter
- 1 VOM (volt-ohm-milliammeter) with alligator clips on the leads, Simpson, 260 series
- 1 Digital VOM meter, Beckman
- 1 Millivolt meter capability, The Beckman meter has this feature
- 1 Pulley puller
- 1 Oil dispenser for oiling motors
- 1 Clamp on wattmeter, TIF Instrument Company
- 1 Thermocouple adapter for checking voltage at a thermocouple see text, Figure 5-7
- 1 Shop vacuum cleaner
- 1 Flue gas analysis kit, complete with draft gage, thermometer and smoke tester, Bacarach
- 1 Water manometer for checking gas pressures
- 1 Oil burner nozzle wrench
- 1 Shallow pan for catching oil drippings
- 1 Set of oil pump checking gages
- 1 3/8" drill motor, battery powered preferred
- 1 Assortment of drill bits to 3/8"
- 1 Water hose
- 1 Medium water bucket
- 1 Steel tape measure, 12 foot

LAB 1   Properties of Gases

Name _____ Date _____ Grade _____

OBJECTIVES: Upon completion of this exercise, you should be able to define the properties of natural, propane and butane gases.

INTRODUCTION: You will use the textbook to fill in exercises on the various properties of gases.

TEXT REFERENCES: Unit 1

TOOLS AND MATERIALS: Pencil and calculator.

SAFETY PRECAUTIONS: None

PROCEDURES

1. A cubic foot of air contains _____% nitrogen and _____% oxygen.
2. 1000 cubic feet of air contains _____ cubic feet of oxygen.
3. A gas burner with an output of 110,000 Btuh is proposed for a particular job. How much air will it take to support perfect combustion? _____ How much air will need to be planned for practical combustion? _____
4. If the above burner were to be used for propane, how much air would be required for perfect combustion? _____ How much air will need to be planned for practical combustion? _____
5. A small building requires that the gas heat appliance have an output of 225,000 Btuh for heating. What furnace input would need to be furnished if the furnace has an operating efficiency of 80%? _____
6. What would be the fuel consumption for the above furnace in cubic feet per hour for natural gas? _____ What would the fuel consumption be for propane in cubic feet per hour? _____
7. What would be the fuel consumption for a natural gas furnace in the above problem if the furnace were to have an efficiency of 95%? _____

SUMMARY STATEMENT: Describe the differences in propane, butane and natural gas.

Lab Manual to Accompany Practical Heating Technology

QUESTIONS

1. Air contains _____% oxygen and _____% nitrogen.

2. When natural gas is taken out of the ground, what must be done to it to prepare it to burn as a fuel?

3. Why are propane and butane called LP gases?

4. Which gas has the most carbon? (propane, butane or natural gas) _____

5. What is the typical manifold pressure for natural gas?_____ LP gas? _____

6. State the specific gravity for the following gases.
   A. Natural _____
   B. Propane _____
   C. Butane _____

7. How many cubic feet per hour would a furnace that is 81% efficient consume while operating on propane? _____, butane? _____, natural gas? _____.

8. What is the ignition temperature for the following gases?
   A. Natural _____
   B. Propane _____
   C. Butane _____

9. Why is perfect combustion not used when setting the air adjustment for a gas burner?

10. Describe the difference in a power gas burner and an atmospheric gas burner.

LAB 2

Gas Burners and Heat Exchangers

Name _____ Date _____ Grade _____

OBJECTIVES: Upon completion of this exercise, you should be able to recognize the differences in the various types of gas burners.

INTRODUCTION: Use the text and a standard furnace to describe the characteristics of the various gas burners and heat exchangers.

TEXT REFERENCES: Units 1 and 2

TOOLS AND MATERIALS: Flat blade and Phillips screwdrivers, 1/4 and 5/16" nut drivers and a flashlight.

SAFETY PRECAUTIONS: Turn off the power and lock it out before starting. Turn off the gas before starting.

PROCEDURES

1. Proceed to a standard forced air gas furnace and turn off the gas, turn off the power and lock it out. NOTE: This exercise is hard to perform on a furnace with force or induced draft configuration.
2. Remove the burner access panel.
3. Remove the flue pipe and draft diverter.
4. What type of burners are used with this furnace? (endshot, slotted port, drilled port, or ribbon)
5. How many burners are there? _____
6. What is the capacity for each burner? _____ Btuh
7. Using the flashlight, examine the heat exchanger and describe how the flue gases reach the draft diverter.

8. What material is the heat exchanger made of? _____
9. Replace all panels with the correct fasteners.

MAINTENANCE OF WORK STATION AND TOOLS: Prepare the furnace for the next students exercise making sure the work station is clean.

SUMMARY STATEMENT: Describe how the furnace exchanges heat from the gas to the air stream.

QUESTIONS

1. What is the purpose of the serpentine type of heat exchanger?

2. What is the purpose of the draft diverter?

3. If a heat exchanger were to rust and patches of rust were to fall on the burner, what could the results be?

4. What type of air adjustment did the burner have on the furnace you worked with?

5. The word "clamshell" applies to what component of a furnace?

6. Why should the combustion gases never be allowed to mix with the air from the conditioned space?

7. What causes a heat exchanger to become damaged?

8. What are the symptoms of a damaged heat exchanger?

9. How can a heat exchanger be examined for damage?

10. Where do the flue gases terminate?

LAB 3 — Gas Furnace Familiarization

Name _____ Date _____ Grade _____

OBJECTIVES: Upon completion of this exercise, you should be able to recognize and state the various components of a typical gas furnace.

INTRODUCTION: You will remove the front panels and possibly the blower and motor for the purpose of identifying the characteristics of all parts of the furnace.

TEXT REFERENCES: Unit 2

TOOLS AND MATERIALS: Straight blade and Phillips screwdrivers, 1/4" and 5/16" nut drivers, a 6" adjustable wrench, Allen wrenches, a flashlight, and a gas furnace.

SAFETY PRECAUTIONS: Turn the power to the furnace off and lock it out before beginning this exercise.

PROCEDURES

1. With the power off, remove the front burner and blower compartment panels.

2. Fan information:

    Motor full-load amperage _____ A
    Type of motor _____
    Diameter of motor _____ in.
    Shaft diameter _____ in.
    Motor rotation (looking at motor shaft) _____
    Fan wheel diameter _____ in.
    Width _____ in.
    Number of motor speeds _____, high rpm _____, low rpm _____

3. Burner information:

    Type of burner _____
    Number of burners _____
    Type of pilot safety _____
    Gas valve voltage _____ V
    Gas valve amperage _____ A
    Gas valve pipe size _____ in.

4. Unit nameplate information:

    Manufacturer _____
    Model number _____
    Serial number _____
    Type of gas _____
    Input capacity _____
    Btu/h Output capacity _____ Btu/h
    Voltage _____ V
    Recommended temperature rise _____ °F
    Control voltage _____ V

5. Heat exchanger information:

    What is the heat exchanger made of? _____ (type of metal)
    Number of burner passages _____, Flu size _____ in.
    Type of heat exchanger? (upflow, downflow or horizontal)

6. Replace all panels with the correct fasteners.

Lab Manual to Accompany Practical Heating Technology

MAINTENANCE OF WORK STATION AND TOOLS: Return all tools to their respective places and be sure the work area is clean.

SUMMARY STATEMENT: Describe the combustion process for the furnace that you worked on.

QUESTIONS:

1. What component transfers the heat from the products of combustion to the room air?

2. What is the typical gas manifold pressure for a natural gas furnace?

3. Which of the following gases requires a 100% gas shutoff? (natural or propane)

4. Why does this gas require a 100% shut off?

5. What is the typical control voltage for a gas furnace?

6. Name 2 advantages for this particular control voltage.

7. What is the purpose of the drip leg in the gas piping located just before a gas appliance?

8. What is the typical line voltage for gas furnaces?

9. What is the purpose of the vent on a gas furnace?

10. What is the purpose of the draft diverter on a gas furnace?

LAB 4 — Identification of the Pilot Safety Feature of a Gas Furnace

Name _____ Date _____ Grade _____

OBJECTIVES: Upon completion of this exercise, you should be able to look at the controls of a typical gas furnace and determine the type of pilot safety features used.

INTRODUCTION: You will remove the panels from a typical gas-burning furnace, examine the control arrangement and describe the type of control that prevents the burner from igniting if the pilot light is out.

TEXT REFERENCES: Units 2 and 3

TOOLS AND MATERIALS: A VOM, straight blade and Phillips screwdrivers, a flashlight, 1/4" and 5/16" nut drivers, a box of matches, and a typical gas-burning furnace.

SAFETY PRECAUTIONS: Turn off power and lock it out. Use caution while working with any gas burning appliance. If there is any question, ask your instructor.

PROCEDURES

1. Select a gas-burning furnace and shut the power off and lock it out.

2. Remove the panel to the burner compartment.

3. Remove the cover from the burner section if there is one.

4. Using the flashlight, examine the sensing element that is in the vicinity of the pilot light and compare it to the text, Unit 2. Follow the sensing tube to its termination point to help identify it.

5. What type of pilot safety feature does this furnace have?

6. If the pilot light is lit, blow it out for the purpose of learning how to relight it.

7. Turn the power on and light the pilot light. You may find directions for the specific procedure in the furnace. If not, ask your instructor.

8. Turn the room thermostat to call for heat. Make sure that the burner ignites.

9. Allow the furnace to run until the fan starts, then turn the room thermostat to off and make sure the burner goes out.

10. Stand by until the fan stops running.

11. Replace all panels with the correct fasteners.

MAINTENANCE OF WORK STATION AND TOOLS: Return all tools to their respective places. Leave the area around the furnace clean.

SUMMARY STATEMENT: Describe the exact sequence used for the pilot safety on this furnace.

QUESTIONS

1. Why is it necessary to have a pilot safety shutoff on a gas furnace?

2. How long could gas enter a heat exchanger if a pilot light were to go out during burner operation with a thermocouple pilot safety?

3. Describe how a thermocouple works.

4. Describe a mercury sensor application.

5. What is the typical heat content of a cubic foot of natural gas?

6. Where does natural gas come from?

7. What is the typical pressure for a natural gas furnace manifold?

8. What component reduces the main pressure for a typical gas furnace?

9. How does the gas company make the consumer aware of a gas leak?

10. How does the gas company charge the consumer for gas consumption?

LAB 5 — Gas Furnace Heat Sensing Elements

Name _____ Date _____ Grade _____

OBJECTIVES: Upon completion of this exercise, you should be able to state the use of the various sensing elements on a gas furnace with a thermocouple pilot safety control.

INTRODUCTION: On a gas furnace with a thermocouple pilot safety device you will locate the fan-limit and pilot safety controls, cause each sensing element to function and record the function.

TEXT REFERENCES: Units 3, 4 and 5

TOOLS AND MATERIALS: A gas furnace (upflow is preferred) with a temperature-operated fan-limit control, straight blade and Phillips screwdrivers, 4" and 6" adjustable wrenches, a millivolt meter, a thermocouple adapter (see text Figure 5–7 for an example), and a temperature tester.

SAFETY PRECAUTIONS: Turn the power to the furnace off before installing the millivolt adapter and placing the thermometer probe.

PROCEDURES

1. With the power off, remove the burner compartment panel.
2. Locate the fan-limit control.
3. Locate the gas valve and the thermocouple connection to the gas valve. Install a thermocouple adapter for taking millivolt readings.
4. Follow the instructions on the furnace and light the pilot light.
5. When the pilot light is burning correctly, fasten the millivolt meter to the adapter and record the millivolt reading. _____ mV
6. With only the pilot light burning, blow out the pilot light and record the voltage of the thermocouple at the time the pilot safety valve drops out. _____ mV
7. Insert a temperature tester lead in the furnace plenum just above the heat exchanger, as close to the limit control location as practical. A small hole may need to be drilled in the duct for this test.
8. Turn on the power and start the furnace. Record the temperature near the fan-limit sensor at 2-minute intervals. Mark the time the fan motor starts:

    2 minutes _____ °F,   4 minutes _____ °F,
    6 minutes _____ °F,   8 minutes _____ °F,
    10 minutes _____ °F, 12 minutes _____ °F

9. When the fan motor starts, shut off the power and disconnect one fan motor lead and tape it for safety. Turn the power back on.
10. Let the burner operate and watch for the limit switch to shut the burner off because there is no air circulation. Record the temperature here. _____ °F
11. Turn the power off and replace the fan wire. Remove the temperature lead from the fan-limit control. Be careful not to burn your hands.
12. Start the fan and allow the furnace to cool off.

MAINTENANCE OF WORK STATION AND TOOLS: Replace all panels and make sure the furnace is left as instructed. Return all tools to their places and make sure the work area is clean.

SUMMARY STATEMENT: Describe the action and principle of a thermocouple.

QUESTIONS

1. What are the circumstances that would cause the limit control to shut the main burner off?

2. Approximately how long does it take the thermocouple to shut the gas supply off when the pilot light goes out?

3. What materials are some thermocouples made of?

4. Is the thermocouple in contact with the pilot flame or the main burner flame?

5. Does the thermocouple open a solenoid valve or hold a solenoid valve open?

6. Name two pilot safety devices other than the thermocouple.

7. What is the sensing element in a typical fan-limit switch made of?

8. Are all fan-starting elements activated by the furnace temperature?

LAB 6: Gas Pressure Testing

Name _____ Date _____ Grade _____

OBJECTIVES: Upon completion of this exercise, you should be able to use a water manometer or pressure gage to check gas manifold pressure.

INTRODUCTION: In this exercise, you will locate the various test ports for checking the manifold gas pressure for a gas burning appliance, furnace or boiler.

TEXT REFERENCES: Unit 4

TOOLS AND MATERIALS: Hex head Allen wrench set, flat blade and Phillips screwdrivers, adjustable wrench, 1/4 and 5/16" nut drivers and method for measuring gas pressure (water manometer preferred for most accurate reading).

SAFETY PRECAUTIONS: Turn off the power and lock it out, then turn off the gas before removing plug at the test port.

PROCEDURES

1. Turn off the power and lock it out.
2. Turn off the gas.
3. Remove the plug from the test port at the burner manifold. NOTE: It may be in the gas valve body.
4. Connect the pressure measuring instrument to the test port and secure the instrument so you can have both hands free.
5. Turn on the power and the gas.
6. Start the furnace, when the gas valve opens to the main burner, pressure should be recorded on the pressure measuring device. If the device is a water manometer, record the difference in the height of the columns. _____ inches of water column
7. If the pressure is not correct, adjust the pressure to the correct pressure for the application.
8. When completed, turn off the furnace and remove the test instrument. Then fasten the plug back in the test port.
9. Replace all panels that were removed.

MAINTENANCE OF WORK STATION AND TOOLS: Replace all tools to their respective places.

SUMMARY STATEMENT: Describe the results of too much pressure on a gas burning appliance.

QUESTIONS

1. Why is it necessary to maintain the correct gas pressure for a gas burning appliance?

2. What device is used to set the gas manifold pressure on a gas burning appliance?

3. What is the typical gas manifold pressure for natural gas? _____, for propane gas? _____

4. What meters the gas to the individual gas burners on a gas burning appliance?

5. What is the gas pressure furnished to the pilot light for a natural gas furnace? _____, for propane gas furnace? _____

6. What is the typical gas pressure in the gas main before the meter at a residence?

7. What pressure is the gas typically metered at for a residence?

8. Why does the gas company furnish gas pressure higher than the manifold pressure?

9. What gas pressure is typically supplied to the piping to the appliances in a residence?

10. What other pressure is now being supplied to some residences and commercial accounts at the appliances?

LAB 7 Orifice Sizing

Name _____ Date _____ Grade _____

OBJECTIVES: Upon completion of this exercise, you should be able to choose the correct orifice size for a gas burning appliance.

INTRODUCTION: You will use the text and the orifice sizing tables to choose the correct size orifices for a gas burning appliance.

TEXT REFERENCES: Unit 4

TOOLS AND MATERIALS: Gas burning appliance, a flashlight and a calculator.

SAFETY PRECAUTIONS: Turn off gas and turn off the power and lock it out while checking furnace data.

PROCEDURES

1. Turn off the power and lock it out.
2. Remove the front from the gas burning appliance so that the nameplate can be observed.
3. Record the input rating of the gas appliance. _____ Btuh
4. Record the number of burners in the appliance. _____
5. Divide the number of burners into the appliance input to determine the capacity per burner and record. _____
6. Using table Figure 4-59 in the text, record the orifice size per burner for natural gas (specific gravity 0.6). _____
7. Using table Figure 4-61 in the text, record the orifice size per burner for propane (specific gravity 1.52). _____
8. Using table Figure 4-60 in the text, record the orifice size for natural gas for this appliance at an altitude of 5000 feet. _____
9. Remove one of the burners from the furnace and record the size of the orifice for that burner. _____
10. Does the orifice in the furnace compare with the calculated orifice size? _____ If not, why?
11. Replace all panels and return the unit to the condition you found it.

MAINTENANCE OF WORK STATION AND TOOLS: Return all tools to their respective places.

SUMMARY STATEMENT: Describe how the orifice meters the gas to the appliance.

QUESTIONS

1. What is the typical manifold pressure for a natural gas burning appliance? _____ A propane burning appliance? _____

2. Is a propane orifice larger or smaller than a natural gas orifice?

3. If a furnace were to have propane connected to it after being operated on natural gas what would the burner characteristics be like if the orifices were not changed?

4. What type of air is induced into the burner mixing tube with the gas stream from the orifice?

5. How would a technician know what size orifice a burner has by looking on the orifice?

6. Suppose an orifice is marked with the number 42, how much capacity would this orifice have using natural gas? _____

7. How much capacity would the above orifice have using propane gas?

8. If the gas manifold pressure were to be above normal for the orifice in the above questions, would the results be?

9. When a gas appliance is converted from natural gas to propane, what should be done about the pilot light orifice?

10. A natural gas boiler is installed at Denver, Colorado at an altitude of 5000 feet. The orifice size shipped in the furnace is size 30, what would the new orifice size be? _____

LAB 8 — Wiring for a Gas Furnace

Name _____ Date _____ Grade _____

OBJECTIVES: Upon completion of this exercise, you should be able to follow the control sequence and wiring, and state the application of the various controls.

INTRODUCTION: You will trace the wiring on a gas furnace and develop a pictorial and a ladder-type diagram for a gas furnace.

TEXT REFERENCES: Unit 3

TOOLS AND MATERIALS: A typical gas furnace with a thermocouple safety pilot, a flashlight, colored pencils, and straight blade and Phillips screwdrivers.

SAFETY PRECAUTIONS: Turn off the power to the furnace before starting this exercise.

PROCEDURES

1. Make sure the power is off and remove the front cover to the furnace.

2. Remove the cover to the junction box where the power enters the furnace.

3. Study the wiring of the furnace. After you feel you understand the wiring, use the space provided to draw a pictorial and a ladder wiring diagram of the complete high- and low-voltage circuit of the gas furnace. Hint: draw each component, showing all terminals, in the approximate location first and then draw the wire connections. DO NOT try to draw the thermocouple circuit.

    PICTORIAL DIAGRAM   LADDER DIAGRAM

4. Replace all covers and panels.

MAINTENANCE OF WORK STATION AND TOOLS: Return all tools and clean the work area.

SUMMARY STATEMENTS: Using the ladder diagram, describe the complete sequence of events required to operate the gas valve and the fan. Start with the thermostat. Explain what happens with each control.

QUESTIONS

1. What is the job of the thermocouple?

2. What is a thermopile? Why are they used?

3. Why shouldn't the thermocouple be tightened down too tightly at the gas valve?

4. Does the limit control shut off power to the transformer or to the gas valve in the furnace used for this exercise?

5. Why is the typical line voltage for a gas furnace 115 volts?

6. Did the furnace in the diagram above have a door switch?

7. If a furnace had a door switch, how would the technician troubleshoot the furnace electrical circuit when the door has to be removed?

8. At how many speeds will the fan motor, previously illustrated, operate?

9. Why is the typical low voltage for a gas furnace 24 volts?

10. Does the furnace in the diagram above have 1 or 2 limit controls?

LAB 9 — Maintenance of Gas Burners

Name _____ Date _____ Grade _____

OBJECTIVES: Upon completion of this exercise, you should be able to perform routine maintenance on an atmospheric gas burner.

INTRODUCTION: You will remove the burners from a gas furnace, perform routine maintenance on the burners, place them back in the furnace, and put the furnace back into operation.

TEXT REFERENCES: Unit 3

TOOLS AND MATERIALS: A gas furnace, straight blade and Phillips screwdrivers, a flashlight, two 6" adjustable wrenches, two 10" pipe wrenches, 1/4" and 5/16" nut drivers, soap bubbles for leak testing, compressed air, goggles and a vacuum cleaner.

SAFETY PRECAUTIONS: Turn off the power and lock it out. Turn off the gas to the furnace off before beginning. Wear safety glasses when using compressed air.

PROCEDURES

1. With the power and gas off, remove the front panel.

2. Disconnect the burner manifold from the gas line.

3. Remove any fasteners that hold the burner or burners in place and remove the burner or burners.

4. Examine the burner ports for rust and clean as needed. Use compressed air and blow the burners out.

5. Vacuum the burner compartment and the manifold area.

6. Remove the draft diverter for access to the top or end of the heat exchanger. Vacuum this area if needed.

7. Using the flashlight, examine the heat exchanger in every area that you can see, looking for cracks, soot and rust.

8. Shake any rust or soot down to the burner compartment and vacuum again. A folded coat hanger or brushes may be used to break loose any debris. You can see how much debris came from the heat exchanger area by vacuuming first and again after cleaning the heat exchanger.

9. Replace the burners in their proper seats and all panels.

10. Turn the gas on and use soap bubbles for leak testing any connection you loosened or are concerned about.

11. Turn the power on and start the furnace. Wait for the burner to settle down and burn correctly. The burner will burn orange for a few minutes because of the dust and particles that were broken loose during the cleaning.

12. When you are satisfied that the burner is operating correctly, shut the furnace off and replace all panels.

Lab Manual to Accompany Practical Heating Technology

MAINTENANCE OF WORK STATION AND TOOLS: Return all tools to their places. Make sure the work area is clean and that the furnace is left as instructed.

SUMMARY STATEMENT: Describe the complete combustion process, including the characteristics of a properly burning flame.

QUESTIONS

1. How many cubic feet of air must be supplied to burn one cubic foot of gas?

2. How many cubic feet of air are normally furnished to support the combustion of one cubic foot of gas?

3. Why is there a difference in the amount of air in questions 1 and 2?

4. Name the three typical types of atmospheric gas burners.

5. What is the difference in an atmospheric burner and a forced-air burner?

6. What is the part of the burner that increases the air velocity to induce the primary air?

7. Where is secondary air induced into the burner?

8. What is the purpose of the draft diverter?

9. What type of air adjustment did the furnace you worked with have?

10. What are the symptoms of a shortage of primary air?

LAB 10

Checking a Gas Furnace for Correct Venting

Name _____ Date _____ Grade _____

OBJECTIVES: Upon completion of this exercise, you should be able to check a typical gas furnace to make sure that it is venting correctly.

INTRODUCTION: You will start up a typical gas furnace and use a match or smoke to determine that the products of combustion are rising up the flue.

TEST REFERENCES: Unit 4

TOOLS AND MATERIALS: A small amount of fiberglass insulation, straight blade and Phillips screwdrivers, 1/4" and 5/16" nut drivers, a flashlight, matches, and a typical operating gas furnace.

SAFETY PRECAUTIONS: Follow the directions with the furnace while lighting the pilot light. Take care while working around the hot burner and flue pipe. The vent pipe gets very hot. It may be 350°F and will burn you.

PROCEDURES

1. Select any gas-burning appliance that is vented through a flue pipe.
2. Remove the cover to the burner and make sure the pilot light is lit. If it is not, follow the appliance instructions and light it.
3. Turn the thermostat up to the point that the main burner lights.
4. While the furnace is heating up to temperature, you may carefully touch the flue pipe to see if it is getting hot.
5. Locate the draft hood. See the text.
6. Place a lit match or candle at the entrance to the draft hood where the room dilution air enters the draft hood. The flame will follow the airflow and should be drawn towards the furnace and up the flue pipe.
7. Shut the furnace off with the thermostat and allow the fan to run and cool the furnace.
8. When the furnace has cooled, place the small amount of fiberglass insulation either at the top or the bottom of the flue pipe to restrict the flow of flue gases.
9. Start the furnace again and place the candle at the entrance to the draft regulator. The flame should sway out towards the room, showing that the appliance flue products are going into the room and not up the flue.
10. Stop the furnace and REMOVE THE FLUE BLOCKAGE.
11. START THE FURNACE AND ONCE AGAIN PLACE THE MATCH FLAME AT THE ENTRANCE TO THE DRAFT DIVERTER AND PROVE THE FURNACE IS VENTING.
12. Replace all panels with the correct fasteners and MAKE SURE THE FLUE IS NOT BLOCKED.

MAINTENANCE OF WORK STATION AND TOOLS: Return all tools to their respective places.

SUMMARY STATEMENT: Describe the function of the draft diverter.

QUESTIONS

1. What is the air called that is drawn into the draft diverter?

2. What gases are contained in the products of complete combustion?

3. What gases are contained in the products of incomplete combustion?

4. Name the component that transfers heat from the products of combustion to the circulated air.

5. Do all gas furnaces have products of combustion?

6. Describe a gas flame that is starved for oxygen.

7. When a gas flame is blowing and lifting off the burner head, what is the likely problem?

8. What does the draft diverter do in the case of a downdraft in the flue pipe?

9. What is the minimum size flue pipe that should be attached to a gas furnace?

10. Is it proper to vent a furnace out the side of a structure without a vertical rise?

LAB 11 — Combustion Analysis of a Gas Furnace

Name _____ Date _____ Grade _____

OBJECTIVES: Upon completion of this exercise, you should be able to perform a combustion analysis on a gas furnace.

INTRODUCTION: You will use a combustion analyzer to analyze a gas furnace for correct combustion.

TEXT REFERENCES: Unit 4

TOOLS AND MATERIALS: Straight blade and Phillips screwdrivers, 1/4" and 5/16" nut drivers, a flue gas analysis kit including thermometer, and an operating gas furnace.

SAFETY PRECAUTIONS: Be very careful working around live electricity. You will be working near a flame and a hot flue gas pipe. Do not burn your hands.

PROCEDURES

1. Remove the panel to the burner section of the furnace.
2. Light the pilot light if needed.
3. Start the furnace burner and watch for ignition.
4. Let the furnace burner burn for 10 minutes and then examine the flame. Compare the flame to the explanation in the text. If the flame needs adjusting, make the adjustments.
5. Following the instructions in the particular flue gas kit that you have, perform a flue gas analysis. Record the results here:
   - Percent carbon dioxide content _____%
   - Flue gas temperature _____°F
   - Efficiency of the burner _____°F
6. Adjust the air shutter for minimum air to the burner and wait 5 minutes.
7. Perform another combustion analysis and record the results here:
   - Percent carbon dioxide content _____%
   - Flue gas temperature _____°F
   - Efficiency of the burner _____°F
8. Adjust the burner air back to the original setting and wait 5 minutes. Perform one last test and record the results here:
   - Percent carbon dioxide content _____%
   - Flue gas temperature _____°F
   - Efficiency of the burner _____°F
9. Turn off the furnace and replace any panels with the correct fasteners.

Lab Manual to Accompany Practical Heating Technology

MAINTENANCE OF WORK STATION AND TOOLS: Return all tools to their respective work stations.

SUMMARY STATEMENT: Describe the difference between perfect combustion and ideal or typical accepted combustion.

QUESTIONS

1. What is CO_2?

2. What is CO?

3. Name the products of complete combustion.

4. What is excess air?

5. Why is perfect combustion never obtained?

6. What is the ideal CO_2 content of flue gas for a gas furnace?

7. How much air is consumed to burn one cubic foot of gas in a typical gas burner?

8. Why must flue gas temperatures in conventional furnaces be kept high?

9. What do yellow tips mean on a gas flame?

10. How is the air adjusted on a typical gas burner?

LAB 12

Determining Airflow (CFM) Using the Air-Temperature-Rise Method

Name _____ Date _____ Grade _____

OBJECTIVES: Upon completion of this exercise, you should be able to determine the amount of air in CFM moving through a gas furnace using the air-temperature-rise method.

INTRODUCTION: You will use a temperature tester and the gas input for a standard efficiency gas furnace to determine the CFM of airflow in a system. A water manometer will be used to verify the correct gas pressure.

TEXT REFERENCES: Unit 4

TOOLS AND MATERIALS: A standard efficiency gas furnace, a temperature tester, and a water manometer.

SAFETY PRECAUTIONS: Shut off the gas before installing the manometer to check the manifold pressure of the gas furnace.

PROCEDURES

1. Make sure the gas is off and install the water manometer on the gas furnace manifold or gas valve pressure tap.

2. Turn on the gas, start the furnace and record the gas pressure at the manifold (between the gas valve and the burners), _____ inches of water column. This should be 3.5 inches of water column for natural gas. If you have a gas other than natural gas, your instructor will have to help you with the correct pressure and Btu content of the gas. If the pressure is not correct, adjust the regulator until it is correct.

3. When you have established the correct gas pressure at the manifold, stop the furnace, shut off the gas and remove the water manometer.

4. Turn on the gas, start the furnace and while it is getting up to temperature, install a thermometer lead in the supply and return ducts. You will get the best reading by placing the supply duct lead several feet from the furnace where the air leaving the furnace will have some distance to mix.

5. Record the temperature difference after the furnace has operated for 15 minutes. _____ °F difference

6. Calculate the furnace gas input based on the nameplate rating in Btuh times 0.80 (0.80 is the decimal equivalent of 80%, the efficiency of a standard gas furnace). For example, a 100,000 Btuh furnace has a typical input of 100,000 × 0.80 = 80,000 Btuh. The remaining heat goes up the flue of the furnace as products of combustion. Your furnace input is _____ nameplate rating × 0.80 = _____ Btuh.

7. Use the following formula to determine the airflow.

$$\text{CFM} = \frac{\text{Total heat input}}{1.1 \times \text{TD (temperature difference)}}$$

8. Stop the furnace and replace all panels.

MAINTENANCE OF WORK STATION AND TOOLS: Return all tools to their places. Make sure the work area is clean and that the furnace is left as you are instructed.

SUMMARY STATEMENT: Describe how heat is transferred from the burning gas to the air leaving the furnace.

QUESTIONS

1. What would the temperature of the leaving air do if the air filter were to become stopped up?

2. What would the flue gas symptoms be if there were too much airflow?

3. What is the Btu heat content of one cubic foot of natural gas?

4. What is the typical high and low temperature difference across a gas furnace? _____ high and _____ low

5. What would cause too much temperature difference?

6. What would be the effects on a furnace and duct system when many of the outlet air registers are shut off to prevent heat in a room?

7. What is the efficiency of a standard gas furnace?

8. What is the efficiency of a high-efficiency gas furnace?

9. How is high efficiency accomplished?

10. What is the manifold pressure of a typical natural gas furnace?

LAB 13

Changing a Gas Valve on a Gas Furnace

Name _____ Date _____ Grade _____

OBJECTIVES: Upon completion of this exercise, you should be able to change a gas valve on a typical gas furnace, using the correct tools and procedures.

INTRODUCTION: You will use the correct wrenches and procedures and change the gas valve on a gas furnace.

TEXT REFERENCES: Unit 2

TOOLS AND MATERIALS: Flat blade and Phillips screwdrivers, needle nose pliers, two 10" adjustable wrenches, 3/8" and 7/16" open end wrenches, two 14" pipe wrenches, soap bubbles for leak checking, VOM, thread seal compatible with natural gas.

SAFETY PRECAUTIONS: Make sure the power is off and locked out, and the gas is off before starting this exercise.

PROCEDURES

1. Turn the power off and lock it out, check it with the VOM.
2. Turn the gas off at the main valve before the unit to be serviced.
3. Remove the door from the furnace. Watch how it removes, so you will know how to replace it.
4. Make a wiring diagram of the wiring to be removed. The wiring may need to be tagged for proper identification. Remove the wires from the gas valve. Use needle nose pliers on the connectors if they are spade type. Do not pull the wires out of their connectors.
5. Look at the gas piping and decide where to take it apart. There may be a pipe union or a flare nut connection to the inlet of the system. Either one may be worked with.
6. Remove the pilot line connections from the gas valve. USE THE CORRECT SIZE END WRENCH AND DO NOT DAMAGE THE FITTINGS.
7. Use the adjustable wrenches for flare fitting connections and pipe wrenches for gas piping connections and disassemble the piping to the gas valve. LOOK FOR SQUARE SHOULDERS ON THE GAS VALVE AND BE SURE TO KEEP THE WRENCHES ON THE SAME SIDE OF THE GAS VALVE. TOO MUCH STRESS CAN EASILY BE APPLIED TO A GAS VALVE BY HAVING ONE WRENCH ON ONE SIDE OF THE VALVE AND THE OTHER ON THE PIPING ON THE OTHER SIDE OF THE VALVE.
8. Remove the valve from the gas manifold. USE CARE NOT TO STRESS THE VALVE OR THE MANIFOLD.
9. While you have the valve out of the system, examine the valve for some of its features, such as where a thermocouple or spark igniter connects. Look for damage.
10. After completely removing the gas valve, treat it like a new one and fasten it back to the gas manifold. BE SURE TO USE THREAD SEAL ON ALL EXTERNAL PIPE THREADS. DO NOT USE EXCESSIVE AMOUNTS OF THREAD SEAL. DO NOT USE THREAD SEAL ON FLARE FITTINGS.
11. Fasten the inlet piping back to the gas valve.
12. Fasten the pilot line back using the correct wrench.

13. Fasten all wiring back using the wiring diagram.

14. Ask your instructor to look at the job and approve, then turn the power and gas on. IMMEDIATELY USE SOAP BUBBLES AND LEAK CHECK THE GAS LINE UP TO THE GAS VALVE.

15. Turn the thermostat to call for heat. WHEN THE BURNER LIGHTS, YOU HAVE GAS TO THE VALVE OUTLET AND THE PILOT LIGHT. LEAK CHECK IMMEDIATELY.

16. When you are satisfied there are no leaks, turn off the furnace.

17. Use a wet cloth and clean any soap bubble residue off the fittings.

MAINTENANCE OF WORK STATION AND TOOLS: Return the work station as the instructor directs you. Return the tools to their respective places.

SUMMARY STATEMENT: Describe what may happen if too much thread seal is used on the fittings to the gas valve.

QUESTIONS

1. Did the gas valve on this system have a square shoulder for holding with a wrench?

2. Did you have any gas leaks after the repair?

3. Why is thread seal used on threaded pipe fittings?

4. Why is thread seal not used on flared fittings?

LAB 14

Changing a Fan Motor on a Gas Furnace

Name _____ Date _____ Grade _____

OBJECTIVES: Upon completion of this exercise, you should be able to change a fan motor on a gas furnace and then restart the furnace.

INTRODUCTION: You will use correct working practices and tools to change a fan motor on a gas furnace.

TEXT REFERENCES: Unit 3

TOOLS AND MATERIALS: Flat blade and Phillips screwdrivers, 1/4" and 5/16" nut drivers, Allen wrenches, small wheel puller, needle nose pliers, VOM, ammeter, 6" adjustable wrench and slip joint pliers.

SAFETY PRECAUTIONS: Make sure the power is off and locked out before starting this exercise. Be sure to discharge any fan capacitors.

PROCEDURES

1. Turn the power off and lock it out, check it with the VOM.

2. Remove the fan compartment door.

3. Make a wiring diagram for the motor wiring so replacement will be easy.

4. Discharge any fan capacitors, and disconnect the motor wires.

5. If the motor is direct drive, remove the blower from the compartment. If the motor is a belt drive, loosen the tension on the belt.

6. If the motor is direct drive, remove the motor from the blower wheel. If the motor is belt drive, remove from the mounting bracket and remove the pulley with puller if needed.

7. Treat this motor as a new motor and replace it in the reverse order that you removed it. Be sure to select correct motor speed. Make sure the oil ports are turned up.

8. Double check the wiring.

9. When all is back together ask your instructor to check the installation.

10. When the installation has been inspected, start the motor and check the amperage. REMEMBER TO CHECK THE AMPERAGE WITH THE FAN DOOR IN PLACE. YOU MAY NEED TO CLAMP THE AMMETER AROUND THE WIRE ENTERING THE FURNACE.

11. Turn the furnace off.

MAINTENANCE OF WORK STATION AND TOOLS: Leave the furnace as your instructor directs you. Return all tools to their places.

SUMMARY STATEMENT: Describe the airflow through a gas furnace.

QUESTIONS

1. What type of fan motor did this furnace have? (shaded pole or permanent split capacitor)

2. Which type of fan motor is the most efficient? (shaded pole or permanent split capacitor)

3. Why would a manufacturer use an inefficient motor when an efficient one is available?

4. What type of motor mount did the motor in this exercise have? (rigid or rubber)

5. What type of bearings did the motor have? (sleeve or ball bearing)

6. Was the motor single or multiple speed?

7. What was the color of the capacitor leads on the unit (if applicable)?

8. What happens to the fan amperage if the leaving airflow is restricted?

9. What happens to the fan amperage if the entering airflow is restricted?

10. Did the motor have oil ports?

LAB 15 — Properties of Oil

Name _____ Date _____ Grade _____

OBJECTIVES: Upon completion of this exercise, you should be able to describe some of the characteristics of fuel oil.

INTRODUCTION: You will use the text to fill in exercises about the various properties of fuel oils.

TEXT REFERENCES: Unit 6

TOOLS AND MATERIALS: Small clear glass container to collect oil sample, small metal plate to burn oil sample, matches, pencil, paper and calculator

SAFETY PRECAUTIONS: Dispose of the sample fuel oil in the proper manner, back in the fuel tank. Move all flammables away from the burn test area. Locate the nearest fire extinguisher.

PROCEDURES

1. Collect some fuel oil in a clear glass container.

2. Describe the color of the oil.

3. Describe how the oil smells and feels to the touch.

4. Apply no more than 3 drops of the fuel oil to a small metal plate that is placed on top of a noncombustible surface. **MOVE ALL COMBUSTIBLE MATERIALS BACK 3 FEET IN EACH DIRECTION.**

5. Use a match to ignite the 3 drops of oil.

6. Describe the flame characteristics of the 3 drops of oil as it burns.

7. How long did the 3 drops of oil burn? _____

8. Describe the smell of the fumes from the burning oil.

9. Return the unused oil to the proper location as directed by your instructor.

MAINTENANCE OF WORK STATION AND TOOLS: Replace all supplies to their respective work stations.

Lab Manual to Accompany Practical Heating Technology

SUMMARY STATEMENT: Describe why the oil flame burned with lazy flame and was slow to ignite.

QUESTIONS

1. A gallon of No. 2 fuel oil contains _____ Btu of heat energy.

2. If a furnace has a nozzle that is marked 0.64 GPH, how much No. 2 fuel oil does it burn per hour? What is the furnace output if the furnace is 70% efficient?

3. What must be done to the liquid fuel oil before it is burned in a modern furnace?

4. Why is fuel oil the preferred petroleum product over gasoline for heating in a home or business?

5. What is the fuel oil pressure in a typical low-pressure fuel oil burner? _____ psig

6. What is the fuel oil pressure in a typical high-pressure fuel oil burner? _____ psig

7. What does the term atomize mean?

8. How many cubic feet of air does it take to burn a pound of fuel oil?

9. Name the by-product of combustion for fuel oil.

10. Describe what the static disc does in a high pressure oil burner.

11. What drives the oil pump in a high pressure oil burner?

12. What is used to ignite the oil in a high pressure oil burner?

LAB 16 — Familiarization with Oil Furnace Components

Name _____ Date _____ Grade _____

OBJECTIVES: Upon completion of this exercise, you should be able to identify the various components of an oil furnace and describe their function.

INTRODUCTION: You will remove, if necessary, study components, and record their characteristics to help you to become familiar with an oil furnace.

TEXT REFERENCES: Unit 8

TOOLS AND MATERIALS: Straight blade and Phillips screwdrivers, two 6" adjustable wrenches, an Allen wrench set, an open end wrench set, a nozzle wrench, rags, and a flashlight.

SAFETY PRECAUTIONS: Turn off the power and oil supply line before starting this exercise.

PROCEDURES

1. After turning off the power and oil supply, remove the front burner and blower compartment panels.

2. Fan information (you may have to remove the fan):

    Motor full-load amperage _____ A   Type of motor _____
    Diameter of motor _____ in.   Shaft diameter _____ in.
    Motor rotation (looking at motor shaft) _____
    Fan wheel diameter _____ in.   Width _____ in.
    Number of motor speeds _____ High rpm _____ Low rpm _____

3. Burner information (burner may have to be removed):

    Nozzle size _____ gpm   Nozzle angle _____ degrees
    Pump speed _____ rpm   Pump motor amperage _____ A
    Number of pump stages _____   One- or two-pipe system _____
    Type of safety control, cad cell or stack switch _____

4. Nameplate information:

    Manufacturer _____   Model number _____
    Serial number _____   Capacity _____ Btuh
    Furnace voltage _____ V   Control voltage _____ V
    Recommended temperature rise, low _____, high _____

5. Is the oil tank above or below the furnace? _____

6. Is there an oil filter in the line leading to the furnace? _____ What is the oil line size? _____

7. Reinstall all components. Have your instructor inspect to ensure that the furnace is in working order.

MAINTENANCE OF WORK STATION AND TOOLS: Replace all furnace panels and leave as you are instructed. Return all tools and materials to their respective places. Clean your work area.

SUMMARY STATEMENT: Describe the complete heating process from the oil supply to the heat transfer method to the air in the conditioned space.

QUESTIONS

1. What is the heat content of a gallon of No. 2 fuel oil?

2. What unit of heat is an oil furnace capacity expressed in?

3. Are oil furnace capacities expressed in output or input?

4. What is the purpose of a two-pipe oil supply system?

5. How is the fan motor started in a typical oil furnace?

6. What is the typical voltage for an oil furnace?

7. What is the typical control voltage for an oil furnace?

8. What is the function of the limit control for an oil furnace?

9. What is the function of the stack switch on an oil furnace?

10. What is the function of a cad cell on an oil furnace?

LAB 17 — Oil Burner Maintenance

Name _____ Date _____ Grade _____

OBJECTIVES: Upon completion of this exercise, you should be able to perform routine maintenance on a typical gun-type oil burner. You will also be able to describe the difference in resistance in a cad cell when it "sees" or does not "see" light.

INTRODUCTION: You will remove a gun-type oil burner from a furnace and remove and inspect the oil nozzle. You will also check and set the electrodes.

TEXT REFERENCES: Unit 10

TOOLS AND MATERIALS: Straight blade and Phillips screwdrivers, a set of open end wrenches (3/8" through 3/4"), two 6" adjustable wrenches, rags, a small shallow pan, a VOM (volt-ohm-milliammeter), a flashlight, and a nozzle wrench.

SAFETY PRECAUTIONS: Shut off the power and oil supply before starting this exercise.

PROCEDURES

1. With the electrical power and oil supply off, disconnect the oil supply line (and return if any).
2. Disconnect the electrical connections to the burner. There should be both high- (115 V) and low-voltage connections.
3. Remove the burner from the furnace and place the burner assembly on a work bench. See Figures 6–16 and 6–19 in the text for an example of a burner assembly.
4. Loosen the fastener that holds the transformer in place. The transformer will usually raise backwards and stop. This exposes the electrode and nozzle assembly.
5. Loosen the oil line connection on the back of the burner. This will allow the nozzle assembly to be removed from the burner. NOTE: This can be removed with the burner in place on the furnace, but we are doing it on a work bench for better visibility.
6. Examine the electrodes and their insulators. The insulators may be carefully removed and cleaned in solvent if needed. They are ceramic and very fragile.
7. Remove the oil nozzle using the nozzle wrench. Inspect and replace or reinstall old nozzle. Follow instructions.
8. Set the electrodes in the correct position, using an electrode gage or text Figure 7–34, as a guide.
9. Replace the nozzle assembly back into the burner assembly.
10. Remove the cad cell (if this unit has a cad cell).
11. Connect your ohmmeter to the cad cell. Turn the cell eye toward a light and record the ohm reading here. _____ Ohms
12. Cover the cell eye and record the ohm reading here. _____ Ohms
13. Replace the cad cell in the burner assembly. Make sure the burner is reassembled correctly before replacing it on the furnace. Check to see that all connections are tight.
14. Return the burner assembly to the furnace and fasten the oil line or lines. Make all electrical connections.

MAINTENANCE OF WORK STATION AND TOOLS: Return all tools to their places. Make sure there is no oil left on the floor or furnace.

SUMMARY STATEMENT: Describe the complete burner assembly, including the static disc in the burner tube and the cad cell.

QUESTIONS

1. Why should you be sure that no oil is left on any part, in addition to it being a possible fire hazard?

2. What is the color of No. 2 fuel oil?

3. What must be done to oil to prepare it to burn?

4. Why is the typical oil burner called a "gun-type" oil burner?

5. What is the typical high pressure nozzle pressure?

6. Is there a strainer in an oil burner nozzle?

7. What is the purpose of the tangential slots in an oil burner nozzle?

8. What provides the pressure for a high-pressure oil burner?

9. What provides ignition for an oil burner?

LAB 18

Electronic Controls for Oil Burners

Name _____ Date _____ Grade _____

OBJECTIVES: Upon completion of this exercise, you should be able to follow the electrical circuit in an oil furnace and check a cad cell for correct continuity.

INTRODUCTION: You will use a VOM (volt-ohm-milliammeter) to follow the line voltage circuit through a primary control and a cad cell on a typical oil furnace.

TEXT REFERENCES: Unit 8

TOOLS AND MATERIALS: A VOM with one lead with an alligator clip and the other lead with a probe, a screwdriver, electrical tape, and an oil furnace that has a cad cell and primary control.

SAFETY PRECAUTIONS: Be sure to turn the power off while removing the primary control from its junction box and performing the cad cell check. Be very careful while making electrical measurements. Make these only under the supervision of your instructor.

PROCEDURES

1. Turn the power off and remove the primary control from its junction box. Suspend it to the side where voltage readings may be taken.

2. Set the voltmeter to the 250-volt scale.

3. Use a lead with an alligator clip and place it on the white wire (neutral).

4. Start the oil furnace and make sure that ignition takes place and that the furnace is fired.

5. Prepare to take voltage readings on the black and orange wires. Remember the black wire is the hot lead feeding the primary and the orange wire is the hot lead leaving the primary to the burner motor and transformer.

6. Using the pointed end of the second probe, check for power at the black terminal and record here. _____ V

7. Check the voltage at the orange terminal and record here. _____ V

8. Turn off the furnace power supply.

9. Remove the cad cell. This is usually accomplished by removing one screw that holds the transformer and raising the transformer to one side.

10. Using the ohm feature of the meter, check the resistance of the cad cell with the eye pointed at the room light source and record it here. _____ Ohms
    Check the resistance with the eye covered with electrical tape. _____ Ohms

11. With the electrical tape still covering the eye, replace the cad cell in its mounting in the furnace and prepare the furnace for firing.

12. Place the alligator clip lead on the white wire. Start the furnace while measuring the voltage at the orange wire. The furnace should run for about 90 seconds before shutting off.

13. Record the running time here. _____ Sec. If you miss the timing the first time, allow 5 minutes for the control to cool and reset the primary. CAUTION: DO NOT RESET THE PRIMARY MORE THAN 3 TIMES OR EXCESS OIL MAY ACCUMULATE IN THE COMBUSTION CHAMBER.

Lab Manual to Accompany Practical Heating Technology

14. Turn the power off, remove the tape from the cad cell, start the furnace and run it for 10 minutes to clear any excess oil from the combustion chamber.

15. Remove one lead from the cad cell and tape it to keep it from touching another circuit. This stimulates an open circuit in the cad cell.

16. Start the furnace while measuring the voltage from the white to the orange wire. Record the voltage here. _____ V

17. Remove the meter leads and place all panels and screws back in their correct order.

MAINTENANCE OF WORK STATION AND TOOLS: Return all tools to their proper places.

SUMMARY STATEMENTS: Describe how the cad cell and the primary control work together to protect the oil furnace from accumulating too much oil in the combustion chamber.

QUESTIONS

1. What material is in a cad cell that allows it to respond to light?

2. What is the response that a cad cell makes to change in light?

3. What would the symptoms be of a dirty cad cell?

LAB 19 — Checking the Fan and Limit Control

Name _____ Date _____ Grade _____

OBJECTIVES: Upon completion of this exercise, you should be able to check a fan and limit switch on an oil furnace for correct action.

INTRODUCTION: You will use a temperature tester probe in the area of the fan and limit control to determine the temperature and then observe the control action.

TEXT REFERENCES: Unit 8

TOOLS AND MATERIALS: Straight blade and Phillips screwdrivers, a temperature tester, 1/4" and 5/16" nut drivers, and a flashlight.

SAFETY PRECAUTIONS: Turn the power off before installing the temperature tester sensor. Be careful not to get burned on hot surfaces.

PROCEDURES

1. With the power off, remove the front cover to the furnace and locate the fan-limit switch.
2. Remove the cover from the fan-limit control and loosen the screws that fasten the control to the furnace. Slide the control from its mounting hole. The wiring should not have to be removed.
3. Slide the temperature tester sensor in next to the control sensor and slide the control back into the furnace. The control does not have to be all the way in, but the sensor should be. Do not pinch the temperature tester lead.
4. Start the furnace and watch the temperature climb. Record the temperature every 2 minutes.

    2 minutes _____ °F   4 minutes _____ °F
    6 minutes _____ °F   8 minutes _____ °F
    10 minutes _____ °F   12 minutes _____ °F

5. Record the temperature when the fan starts. _____ °F
6. When the fan starts, shut the furnace off using the system switch. The furnace will not have time to cool.
7. Disconnect a wire from the fan motor so the fan will not start.
8. Start the furnace again and record the temperature when the limit switch stops the burner pump and fan. _____ °F
9. Shut the power off and reconnect the fan wire.
10. Turn the power on and the fan should start and cool the furnace.
11. Wait until the fan cools the furnace to the point that the fan stops. Then remove the temperature lead from the area of the fan-limit control.
12. Replace the fan-limit control and the front panel.
13. Start the furnace again to make sure it will start for the next exercise.

MAINTENANCE OF WORK STATION AND TOOLS: Return all tools to their proper places and make sure your work area is clean.

SUMMARY STATEMENT: Describe what happens within the fan-limit control from the time the stack starts to heat until it cools to the point where the fan shuts off.

QUESTIONS

1. State a common cause for a limit switch to stop a forced-air furnace.

2. Describe the sensor on the fan-limit control that senses heat.

3. What would happen if a technician wired around a limit switch and the fan became defective and would not run?

4. What would the symptoms be on an oil furnace when the fan setting on the fan-limit control was set too low?

5. Describe the location of the fan-limit control on the furnace you worked with.

6. How many wires are attached to the fan-limit control on the furnace you worked with?

7. What circuit does the limit control break when there is an overheat condition?

LAB 20

Adjusting an Oil Burner Pump Pressure

Name _____ Date _____ Grade _____

OBJECTIVES: Upon completion of this exercise, you should be able to install the gages on an oil pump and adjust it to the correct pressure.

INTRODUCTION: You will install a pressure gage on the oil pump discharge and a compound gage (that will read into a vacuum) on the suction side of the oil pump. You will then start the pump and adjust the discharge pressure to the correct pressure.

TEXT REFERENCES: Unit 9

TOOLS AND MATERIALS: Straight blade and Phillips screwdrivers, a vacuum or compound gage, a pressure gage (150 psig), any adapters to adapt gages to an oil pump, a shallow pan to catch oil drippings, 2 adjustable wrenches, a thread seal, a set of Allen wrenches, and an operating oil burning furnace.

SAFETY PRECAUTIONS: Do not spill any fuel oil on the floor; use a pan under the pump while connecting and disconnecting gages.

PROCEDURES

1. Shut off the power to the furnace.
2. Shut the fuel supply valve off if the supply tank is above the oil pump.
3. Remove the gage plug for the discharge pressure. Usually this is the one closest to the small oil line going to the nozzle. Connect the high-pressure gage here. See text, Figure 9–18 for an example.
4. Remove the gage plug for the suction side of the pump. This is usually located next to the inlet piping on the pump.
5. Remove the pressure-regulating screw protective cap and insert the correct Allen wrench in the slot.
6. Open the fuel valve, start the furnace and quickly make sure that there are no oil leaks. If there are, shut the furnace off and repair them. IF THERE ARE NO LEAKS, MAKE SURE THAT THE OIL HAS IGNITED AND IS BURNING.
7. Record the pressures on the gages here:

    * Inlet pressure   _____ psig or in. Hg
    * Outlet pressure _____ psig

8. Adjust the outlet pressure to 90 psig (this is 10 psig below the recommended) then back to 100 psig. Count the turns per 10 psig.
9. Adjust the outlet pressure to 110 psig (this is 10 psig above the recommended pressure) then back to 100 psig. Count the turns per 10 psig.
10. Shut the furnace off and remove the gages.
11. Wipe any oil away from the fittings and dispose of any oil in the pan as your instructor advises.
12. Replace all panels with the correct fasteners.

MAINTENANCE OF WORK STATION AND TOOLS: Make sure the work station is clean and return all tools to their proper places.

SUMMARY STATEMENT: Describe how the oil burner pump responded to adjustment — approximately how many turns per 10 psig change.

QUESTIONS

1. What is the typical recommended oil burner nozzle pressure?

2. Why should fuel oil not be spilled?

3. What was the size of the pressure tap in the oil pump for the high-pressure gage?

4. What was the size of the pressure tap in the oil pump for the low-pressure gage?

5. How many Btu are in a gallon of fuel oil?

6. What is the typical efficiency for an oil furnace?

7. When is a 2-pipe system necessary for an oil burner?

8. How far vertically can an oil burner pump without an auxiliary pump?

9. Name the 2 places that oil is filtered before it is burned.

10. What is the name of the slots in an oil burner nozzle?

LAB 21 — Combustion Analysis of an Oil Burner

Name _____ Date _____ Grade _____

OBJECTIVES: Upon completion of this exercise, you should be able to perform a draft check, a smoke test and analyze an oil burner for combustion efficiency.

INTRODUCTION: You will use a draft gage to measure the draft over the fire, a smoke tester to measure the smoke in the flue, and use a combustion analyzer to perform a combustion analysis on an oil-burning system.

TEXT REFERENCES: Unit 9

TOOLS AND MATERIALS: A drill motor, a 1/4" drill bit, 1/4" and 5/16" nut drivers, straight blade and Phillips screwdrivers, a draft gage, a smoke test kit, a combustion analysis kit, and an operating oil-burning furnace.

SAFETY PRECAUTIONS: The flue pipe of an oil furnace is very hot while running; do not touch it.

PROCEDURES

1. If holes are not already available, drill a 1/4" hole in the flue pipe at least 6" before the draft regulator and one in the burner inspection door.

2. Insert the thermometer from the analysis kit in the hole in the flue pipe.

3. Start the furnace and determine that the oil ignites.

4. Wait for 10 minutes or until the thermometer in the flue stops rising; then use the draft gage and measure the draft above the fire, at the inspection door. Record the reading here. _____ in.

5. Remove the thermometer from the flue pipe and perform a smoke test, following the instructions in the kit. Record the smoke test number here. _____

6. Perform a combustion analysis at the 1/4" hole in the flue pipe. Record the CO_2 reading here. _____

7. Record the combustion efficiency here. _____ % efficient

8. Turn the furnace off and remove the instruments.

9. Replace all panels with the correct fasteners.

MAINTENANCE OF WORK STATION AND TOOLS: Return all tools to their proper places.

SUMMARY STATEMENTS: Describe the efficiency of this oil furnace. If it was running below 70% efficient, explain the probable causes.

QUESTIONS

1. What may be the cause of a high stack temperature?

2. How would you prove a furnace was stopped up with soot in the heat exchanger?

3. What is the function of the heat exchanger?

4. What is the function of the combustion chamber?

5. What is the purpose of the restrictor (static disc) in the blast tube?

6. What is the difference between a hollow-core and a solid-core oil nozzle?

7. What is the purpose of the tangential slots in the nozzle?

8. What must be done to oil to prepare it to burn?

9. What would be the result of a situation where a customer reset the primary control many times before the service person arrived?

10. What are the symptoms of an oil forced-air furnace with a hole in the heat exchanger?

LAB 22 — Oil Furnace, Changing the Nozzle

Name _____ Date _____ Grade _____

OBJECTIVES: Upon completion of this exercise, you should be able to correctly change an oil burner nozzle using the correct tools and procedures.

INTRODUCTION: You will use the correct tools and procedures to change an oil burner nozzle. After the repair, you will start the furnace.

TEXT REFERENCES: Unit 7

TOOLS AND MATERIALS: Flat blade and Phillips screwdrivers, Allen wrenches, two 6" adjustable wrenches, a small pan, rags and a nozzle wrench.

SAFETY PRECAUTIONS: Be sure the power is off before removing any parts.

PROCEDURES

1. Start the furnace with the old nozzle to be sure it runs.
2. Turn the furnace off and shut off the power.
3. Remove the small fuel line going into the burner from the pump.
4. Remove the nut that holds the fuel line to the burner.
5. Remove or raise the transformer.
6. Remove the burner assembly out the transformer opening. Be careful not to spill oil on the floor. Catch oil in the pan.
7. With the nozzle out, drain the oil into the pan. Set the oil line end of the assembly in the pan and use the nozzle wrench to remove the old nozzle. Oil will drain out the inlet end when the nozzle is loosened.
8. Check the new nozzle against the old nozzle for size, spray pattern and angle. Install the new nozzle if it is correct.
9. Turn the burner assembly up on the nozzle end and add oil to the inlet pipe. Do not allow the nozzle to touch anything or it may become damaged or clogged.
10. Hold your thumb over the inlet end and reinstall the burner assembly, trying not to lose oil from the inlet pipe.
11. Fasten the inlet pipe to the housing and then fasten the small line back to the inlet pipe.
12. Wipe away any spilled oil from the burner assembly and floor.
13. Have the instructor inspect the job.
14. With the instructor's permission, start the unit. You should run an efficiency test after a nozzle change. Ask your instructor.

MAINTENANCE OF WORK STATION AND TOOLS: Return the furnace to the condition your instructor advises you and return all tools to their places. Do not leave any loose oil or wet surfaces on the furnace.

SUMMARY STATEMENT: Describe the action of the burner nozzle in conjunction with the burner blast tube.

QUESTIONS

1. Describe the tangential slots in the burner nozzle.

2. Describe what would happen if you did not fill the burner inlet pipe with oil after a nozzle change.

3. Were the electrodes in good shape on the burner you worked on?

4. When you disconnected the transformer, did it come completely off, or did it have hinges?

5. What size tube ran from the oil pump to the burner assembly?

6. What was the burner size in gallons per minute?

7. What was the angle of spray?

8. What was the spray pattern?

9. Was this furnace a 1- or 2-pipe system?

10. Describe the difference in a 1- and 2-pipe system.

LAB 23 — Oil Furnace, Changing an Oil Pump

Name _____ Date _____ Grade _____

OBJECTIVES: Upon completion of this exercise, you should be able to change an oil pump on a gun-type oil burner.

INTRODUCTION: You will use good workmanship and recommended practices to change an oil pump on an oil furnace. You will then start the furnace up and check it for correct operation.

TEXT REFERENCES: Unit 7

TOOLS AND MATERIALS: Flat blade and Phillips screwdrivers, two 6" adjustable wrenches, flue gas analysis kit, a small set of socket wrenches, Allen wrenches, an oil gage, a pan and rags.

SAFETY PRECAUTIONS: Make sure the power is off and locked out before removing any parts.

PROCEDURES

1. If the furnace is operating, start it and make sure it is firing correctly.

2. Stop the furnace, turn off and lock out the power.

3. Place the pan under the oil pump and remove the oil lines, one or two, depending on the system.

4. Remove the oil pump supply line to the burner. Catch any oil in the pan.

5. Loosen the ignition transformer and raise it out of the way.

6. Using an Allen wrench, loosen the coupling on the oil pump end.

7. Remove the bolts holding the oil pump on the burner.

8. With the oil pump off, you have the same as a new oil pump. Your instructor may want you to reinstall it. With a new oil pump, you would need to check the bypass plug and the connections.

9. Before installing the new pump, fill it with oil so it will be fully primed for startup. It will then pump from the beginning.

10. Install the new pump (or replace the old one) in the reverse order of removing the old pump. If you have questions, ask your instructor.

11. Wipe up any oil that may be around the burner or furnace and move the pan that may have oil in the bottom back from the furnace.

12. Install a gage on the leaving oil pressure tap.

13. When the installation is complete, ask your instructor to approve the installation.

14. Start the furnace and allow it to reach full operating temperature. Make sure the oil pressure is correct, then run a flue gas analysis.

15. Turn the furnace off, allowing the fan to run and cool the furnace.

MAINTENANCE OF WORK STATION AND TOOLS: Clean all oil from around the furnace. Replace all tools to their places.

SUMMARY STATEMENT: Describe the combustion process in the combustion chamber. Tell what the products of combustion are.

QUESTIONS

1. What was the oil pressure with the new pump?

2. Was the furnace operating at peak efficiency with the new pump after startup?

3. What happens if the oil from the nozzle reaches the far side of the combustion chamber?

4. What would the result be if the oil pressure were too high?

5. How low can the oil tank be below the oil burner before the oil pump cannot pump the oil to the burner?

6. What is the typical combustion chamber made of?

7. What is the result of an oil nozzle that drips at the end of the cycle?

8. What can cause a dripping nozzle?

9. What can be done to correct a dripping nozzle?

10. Does fuel oil burn as a liquid?

LAB 24

Making Voltage and Amperage Readings with a VOM

Name _____ Date _____ Grade _____

OBJECTIVES: Upon completion of this exercise, you should be able to make voltage and amperage readings on actual operating equipment using a VOM. You will be able to do this under the supervision of your instructor.

INTRODUCTION: You will be using a VOM to make voltage readings and a clamp-on type ammeter to make AC amperage readings. Your ammeter may be an attachment to your VOM or it may be designed to use independently.

TEXT REFERENCES: Unit 11

TOOLS AND MATERIALS: A VOM with insulated alligator clip test leads, a clamp-on type ammeter, 1/4" and 5/16" nut drivers, and a straight blade screwdriver.

SAFETY PRECAUTIONS: Working around live electricity can be very hazardous. Use a meter with insulated alligator clips on the ends of the leads. Make all meter connections with the POWER OFF. Have your instructor inspect and approve all connections before turning the power on. Your instructor should have given you thorough instruction in using the VOM and clamp-on ammeter. The meters will vary in design and you will need specific instruction in the use of each. The scales on your meter may differ from those indicated in this exercise.

PROCEDURES

Making Voltage Readings

1. With the power off, locate the control voltage transformer in a heating unit.

2. Determine the output (24 V) terminals.

3. Turn the function switch to AC and the range switch to 50 V on your VOM. The black test lead should be in the common (-) jack and the red test lead in the (+) jack in your meter.

4. Connect the insulated alligator clips to the transformer output (24 V) terminals.

5. Have your instructor check the connections. Turn the power on. Record the AC reading: _____

6. Turn off the power.

7. Remove the leads.

8. You will now measure the voltage at the input side of the transformer. The meter function switch should be set at AC. Set the range switch at 250 V.

9. With the power off, fasten the meter test leads with alligator clips to the two terminals at the input side of the control transformer.

Lab Manual to Accompany Practical Heating Technology

10. Ask your instructor to check the connections. With approval turn the power on. Record the AC voltage: _____

11. Turn the power off.

12. Remove the leads.

13. Ask your instructor to show you the location of a 230 V power supply.

14. Set the meter function switch at AC and the range switch at 500 V. (Always select a setting higher than the anticipated voltage.)

15. Connect the meter lead alligator clips to the terminals.

16. Have your instructor inspect the connections and with approval turn the power on. Record the voltage reading: _____

17. Turn power off.

18. Remove the leads.

Making Amperage Readings

1. With the power off, place the clamp-on ammeter jaws around a wire to the input of a fan motor or other appropriate load.

2. Have your instructor check your setup including any necessary settings on your meter.

3. With approval turn the power on and start the fan. Record the amperage reading: _____

4. Turn the power off.

5. Remove the meter.

6. With the power off, disconnect one lead from the output of the 24-V control transformer. Wrap a small wire ten times around one jaw of the ammeter. Connect one end of this wire to the terminal where you just removed the wire. Connect the other end to the wire you removed. See text Figure 13-34.

7. Ask your instructor to check your setup including any settings you may need to make on your meter. With the instructor's approval turn the power on. Record your amperage reading: _____ Your reading will actually be ten times the actual amperage because of the ten loops of wire around the jaws of the ammeter. Divide your reading by ten and record the actual amperage: _____

8. Turn off the power. Remove the wire and ammeter. Connect the wire back to the transformer terminal as you originally found it.

MAINTENANCE OF WORK STATION AND TOOLS: Replace all panels on equipment with correct fasteners. Return all meters and tools to their proper places.

LAB 24 — Making Voltage and Amperage Readings with a VOM *(continued)*

Name _____ Date _____ Grade _____

SUMMARY STATEMENTS: Describe in your own words how you would make a voltage reading in the 230-V range.

Describe how you would use a clamp-on ammeter to make an AC amperage reading.

QUESTIONS

1. Why is alternating current rather than direct current normally used in most installations?

2. Describe why the voltage is different from the input to the output of a transformer.

3. Is a typical control transformer a step-up or step-down transformer?

4. Describe a VOM.

5. What are typical settings for the function switch on a VOM?

6. What is the range switch used for on a VOM?

7. Explain how the magnetic field can be increased around an iron core.

8. Explain how a solenoid switch works.

9. Describe inductance.

10. When measuring voltage, is the meter connected in series or parallel?

LAB 25

Measuring Resistance and Using Ohm's Law

Name _____ Date _____ Grade _____

OBJECTIVES: Upon completion of this exercise, you should be able to make resistance readings using the ohmmeter features of a VOM and you will determine the amperage, voltage or resistance of a circuit using Ohm's Law.

INTRODUCTION: You will be using a VOM to make resistance readings and a VOM to check your calculations using Ohm's Law.

TEXT REFERENCES: Unit 11

TOOLS AND MATERIALS: A VOM with insulated alligator clip test leads, a clamp-on type ammeter, 1/4" and 5/16" nut drivers, and a straight-blade screwdriver.

SAFETY PRECAUTIONS: Use a meter with insulated alligator clips on the ends of the leads. Make all meter connections with the power off. Have your instructor inspect all connections before turning the power on. Your instructor should have given you thorough instruction in using all features of your meter.

PROCEDURES

1. Make the zero ohms adjustment on the VOM in the following manner: Set the function switch to either −DC or +DC. Turn the range switch to the desired ohms range. Probably RX100 will be satisfactory for this exercise. Connect the two test leads together and rotate the zero ohms control until the pointer indicates zero ohms. (Paragraph 11.19 in your text will indicate the ohm range for each range selection.)

2. With the power off and locked out, disconnect one lead to an electric furnace heating element.

3. Clip one lead of the meter to each terminal of the element. Do not turn the power on. Read and record the ohms resistance. _____ Disconnect the meter leads.

4. Connect the heating element back into its original circuit. Place the clamp-on ammeter jaws around one lead to the heating element.

5. Have your instructor inspect and approve your setup. Turn the power on and quickly read the amperage. Record the amperage: _____ You must take the reading quickly because as the element heats up the resistance increases. Turn the power off.

6. Turn the function switch on the meter to AC and set the range switch to 500 V.

7. With the power off, connect a meter lead to each terminal on the heating element.

8. Have your instructor inspect and approve your setup. Turn the power on. Record the voltage: _____ Turn the power off.

9. Check your readings by using Ohm's Law. Paragraph 11.11 in your text explains Ohm's Law. To check the amperage reading use the following formula: $I = E/R$

10. Substitute the voltage and resistance reading from your notes. Divide the voltage by the resistance. Your answer should be close to your amperage reading.

MAINTENANCE OF WORK STATION AND TOOLS: Remove test leads from your meter and coil them neatly or follow your instructor's directions. Replace any panels. Return all tools and equipment to their proper place.

SUMMARY STATEMENTS: Describe the process of zeroing the ohms on a meter and measuring the resistance of a component.

QUESTIONS

1. Describe the difference between a step-up and step-down transformer.

2. What do the letters VOM stand for? What does a typical VOM measure?

3. Describe how to make the zero ohms adjustment on a VOM.

4. What is inductance?

5. Describe how you would use a typical clamp-on ammeter to measure the amperes in a circuit.

6. Explain how a larger diameter wire can safely carry more current than a smaller size.

7. What can happen if a wire too small in diameter is used for a particular installation?

8. How can circuits be protected from current overloads?

9. What are two methods used in most circuit breakers to protect electrical circuits?

LAB 26
Checking the Accuracy or Calibration of Electrical Instruments

Name _____ Date _____ Grade _____

OBJECTIVES: Upon completion of this exercise, you should be able to check the accuracy of and calibrate an ohmmeter, an ammeter, and a voltmeter.

INTRODUCTION: You will use field and bench techniques to verify the accuracy of and calibrate an ohmmeter, a voltmeter, and an ammeter so that you will have confidence when using them. Reference points will be used that should be readily available to you.

TEXT REFERENCES: None

TOOLS AND MATERIALS: A calculator, a resistance heater (an old duct heater will do), several resistors of known value (gold band electronic resistors), an ammeter, a voltmeter, a volt-ohm-milliammeter (all meters with insulated alligator clip leads), a 24-volt, 115-volt, and 230-volt power source and various live components in a system available in your lab.

SAFETY PRECAUTIONS: You will be working with live electrical circuits. Caution must be used while taking readings. Do not touch the system while the power is on. Working around live electrical circuits will probably be the most hazardous part of your job as a heating technician. At this point you probably are not used to working around live electricity. You should perform the following procedures only after you have had proper instruction and only under close personal supervision by your instructor. You must follow the procedures as they are written here as well as the procedures given by your instructor.

PROCEDURES

Ohmmeter Test

1. Turn the ohmmeter selector switch to the correct range for one of the known resistors. For example, if the resistor is 1500 Ohms, the R × 1.00 scale will be correct for most meters because it will cause the needle to read about mid-scale. Your instructor will show you the proper scale for the resistors he gives you to test. Place the alligator clip ends together and adjust the meter to 0 using the 0 adjust knob.

   Using several different resistors of known value, record the resistances:

   Meter range setting, R × _____ Resistor value _____ Ohms
   Meter reading _____ Ohms   Difference in reading _____ Ohms

   Meter range setting, R × _____ Resistor value _____ Ohms
   Meter reading _____ Ohms   Difference in reading _____ Ohms

   Meter range setting, R × _____ Resistor value _____ Ohms
   Meter reading _____ Ohms   Difference in reading _____ Ohms

   Meter range settings, R × _____ Resistor value _____ Ohms
   Meter reading _____ Ohms   Difference in reading _____ Ohms

   Meter range setting, R × _____ Resistor value _____ Ohms
   Meter reading _____ Ohms   Difference in reading _____ Ohms

Voltmeter Test

In this test you will use the best voltmeter you have and compare it to others. Your instructor will make provision for you to take voltage readings across typical low-voltage and high-voltage components. You must perform these tests in the following manner and follow any additional safety precautions from your instructor.

- Use only meter leads with insulated alligator clips.

- You and your instructor should verify that all power to the unit you are working on is turned off.

- Fasten alligator clips to terminals indicated by your instructor.

- Make sure that all range selections have been set properly on your meter.

When your instructor has approved all connections, turn the power on and record the measurement.

1. Take a volt reading of both a high- and low-voltage source using your best meter (we will call it a STANDARD). Your instructor will show you the appropriate terminals. Record the voltages here:

    Low voltage reading _____ V   High voltage reading _____ V

2. Use your other meters such as the VOM meter and the ammeter's voltmeter feature and compare their readings to the standard meter. Record the following:

    STANDARD: low V reading _____ V   high V reading _____ V
    VOM: low V reading _____ V   high V reading _____ V
    Volt scale on
        Ammeter: low V reading _____ V   high V reading _____ V

    Record any differences in the readings here.

    VOM reads: _____ V high, or _____ V low
    Ammeter reads: _____ V high, or _____ V low

Ammeter Test

(Follow all procedures indicated in the voltmeter test.)

1. WITH THE POWER OFF, connect the electric resistance heater to the power supply.

2. Fasten the leads of the most accurate voltmeter to the power supply directly at the heater.

3. Take an ohm reading WITH THE POWER OFF. _____ Ohms

4. Clamp the ammeter around one of the conductors leading to the heater.

5. HAVE YOUR INSTRUCTOR CHECK THE CONNECTIONS and then turn the power on long enough to record the following. (The reason for not leaving the power on for a long time is that the resistance of the heater will change as it heats and this will change the ampere reading.)

    Volt reading _____ V   Ampere reading _____ A

LAB 26

Checking the Accuracy or Calibration of Electrical Instruments *(continued)*

Name _____ Date _____ Grade _____

6. Use a bench meter if possible to compare to the clamp-on ammeter.

   Bench meter reading _____ A   Clamp-on ammeter reading _____ A
   Difference in the two _____ A

7. Double the conductor through the ammeter and watch the reading double, see figure below.

[Figure: Clamp-on ammeter reading 45.5 AMPERES; bench meter reading 22.8 AMPERES; 228 VOLTS 5000 WATTS ELECTRIC HEATER]

MAINTENANCE OF WORK STATION AND TOOLS: Put each meter back in its case with any instruments that it may have. Turn all electrical switches off. Return all tools and equipment to their respective places.

SUMMARY STATEMENT: Describe the accuracy of each meter that you tested.

QUESTION

1. Describe an analog meter.

2. Describe a digital meter.

3. Which of the above meters will usually take the most physical abuse?

4. Why does the amperage change on an electrical heater when it gets hot?

5. What does the gold band mean on a high-quality carbon resistor used for electronics?

6. What should the resistance be in a set of ohmmeter leads when the probes are touched together?

7. Why must you adjust an ohmmeter to 0 with the leads touching each time you change scales?

8. What should be done with a meter with a very slight error?

9. What should be done with a meter that has a large error and cannot be calibrated?

10. Why must you be sure that your meters and instruments are reading correctly and in calibration?

LAB 27 — Electric Furnace Familiarization

Name _____ Date _____ Grade _____

OBJECTIVES: Upon completion of this exercise, you should be familiar with the components in an electric heating system and be able to list the specifications for these components.

INTRODUCTION: You will inspect all of the components in an electric furnace and read specifications off the component nameplates or from available literature.

TEXT REFERENCES: Unit 12

TOOLS AND MATERIALS: Straight blade and Phillips screwdrivers, 1/4" and 5/16" nut drivers, a VOM, temperature tester, calculator, wire size chart, and a complete electric furnace.

SAFETY PRECAUTIONS: Turn the power off to the heating unit before beginning your inspection. After turning the power off, check the line voltage with your VOM to make sure that it is turned off.

PROCEDURES

1. Locate each of the following components. Place a check mark after the name when you have located it.

    - Thermostat   _____
    - Heat anticipator   _____
    - Control device(s) for energizing heating elements (contactor or sequencer) _____
    - Heating elements   _____
    - Fan motor and fan   _____

2. Provide the following information from available literature or component nameplates:

    - Fan motor horsepower   _____
    - Fan motor shaft size   _____
    - Capacitor rating of fan motor (if any)   _____
    - Fan motor FLA   _____
    - Fan motor LRA   _____
    - Fan motor operating voltage   _____
    - Fan diameter   _____
    - Number of heating elements   _____
    - Amperage draw of each heating element   _____
    - KW rating of each heating element   _____
    - Resistance of each heating element   _____
    - Current draw in the control circuit   _____
    - Setting of the heat anticipator   _____
    - Minimum wire size each heating element   _____
    - Btu capacity of each heating element   _____
    - Minimum wire size to furnace   _____
    - Minimum wire size to fan motor   _____

MAINTENANCE OF WORK STATION AND TOOLS: Replace all panels on furnace or leave as you are instructed. Return all tools, VOM and other materials to their proper locations.

SUMMARY STATEMENT: Make a sketch of an electric furnace indicating the probable location of each component.

QUESTIONS

1. Electric heat is rated in which of the following terms? (kW, Btu/h or gal/hr)

2. The individual heater wire size is dependent on which of the following? (voltage, amperage, kW or horsepower)

3. Electric heat is supposed to glow red hot in the air stream. (yes or no)

4. When a fuse blows several times, a larger fuse should be installed. (yes or no)

5. What control component starts the fan on a typical electric heat furnace?

6. When the furnace air filter is stopped up, which of the following controls would cut off the electric heat first? (the limit switch or the fuse link)

7. Which instrument is used to check the current draw in the electric heat circuit? (ohmmeter, voltmeter or ammeter)

8. Is electric heat normally less expensive or more expensive to operate than gas heat?

LAB 28

Determining Airflow (CFM) by Using the Air Temperature Rise

Name _____ Date _____ Grade _____

OBJECTIVES: Upon completion of this exercise, you should be able to calculate the quantity of airflow in cubic feet per minute (CFM) in an electric furnace by measuring the air temperature rise, current and voltage.

INTRODUCTION: You will measure the air temperature at the electric furnace supply and return ducts and the voltage and amperage at the main electrical supply. By using the formula furnished, you will calculate the cubic feet per minute supplied by the furnace. The power supply must be single phase.

TEXT REFERENCES: Unit 12

TOOLS AND MATERIALS: A VOM, an ammeter, a wattmeter (if available), temperature tester, calculator, Phillips and straight blade screwdrivers, 1/4" and 5/16" nut drivers, and an operating electric furnace.

SAFETY PRECAUTIONS: Use proper procedures in making electrical measurements. Your instructor should check all measurement setups before you turn the power on. Use VOM leads with alligator clips. Keep your hands away from all moving or rotating parts.

PROCEDURES

1. Under the supervision of your instructor and with the power off, connect the ammeter and voltmeter where they will measure the voltage and total current supplied to the electric furnace.

2. Turn the unit on. Wait for all elements to be energized. Record the current and voltage below:

    Current _____   Voltage _____
    Current × Volts = Watts
    _____ × _____ = _____ Watts

3. Convert watts to Btu per hour, Btu/h. This is the input. Watts × 3.413 = Btu/h
    _____ Watts × 3.413 = _____ Btu/h

4. Find the temperature difference in the supply and return duct, TD. Place a temperature sensor in the supply duct (around the first bend in the duct to prevent radiant heat from the elements from hitting the probe) and a temperature sensor in the return duct. Record the temperatures below:

    Supply _____ °F − Return _____ °F = _____ °F TD

5. To calculate the cubic feet per minute (CFM) of airflow use the following formula and procedures:

    $$\text{CFM airflow} = \frac{\text{Total heat input (step 3)}}{1.1 \times \text{TD (step 4)}}$$

    $$\text{CFM} = \frac{\text{Input } \underline{\qquad} \text{ Btu/h}}{1.1 \times \underline{\qquad} \text{°F, TD (temperature difference)}}$$

6. If a wattmeter is available you can measure the power of the circuit to check your voltage and current measurements. There will be a slight difference because the motor is an inductive load and does not calculate perfectly using an ammeter and voltmeter to measure power. Do not measure the power until your instructor has checked your setup.

7. Replace all panels on the unit with the correct fasteners.

MAINTENANCE OF WORK STATION AND TOOLS: Return all tools and instruments to their proper places. Leave the furnace as you are instructed.

SUMMARY STATEMENT: Describe why it is necessary for a furnace to provide the correct CFM.

QUESTIONS

1. The kW rating of an electric heater is the _____. (temperature range, power rating, voltage rating or name of the power company)

2. The Btu rating is a rating of the _____. (time, wire size, basis for power company charges or heating capacity)

3. The factor for converting watts to Btu is _____.

4. A kW is _____. (1000 watts, 1200 watts, 500 watts, 10,000 watts)

5. Compared to a gas furnace, the typical air temperature rise across an electric furnace is normally considered _____. (high or low)

6. The air temperature at the outlet grilles of an electric heating system is considered _____. (hot or warm)

7. The circulating fan motor normally runs on _____ (high or low) speed in the heating mode.

LAB 29

Setting the Heat Anticipator for an Electric Furnace

Name _____ Date _____ Grade _____

OBJECTIVES: Upon completion of this exercise, you should be able to take a current reading with a "10-wrap loop" and properly adjust a heat anticipator in an electric furnace low-voltage control circuit.

INTRODUCTION: You will be working with a single-stage and a multistage electric heating furnace, taking current readings in the low-voltage circuit and checking or adjusting the heat anticipator.

TEXT REFERENCES: Text Figure 13-34

TOOLS AND MATERIALS: Phillips and straight blade screwdrivers, 1/4" and 5/16" nut drivers, a clamp-on ammeter, single strand of thermostat wire, a single-stage electric furnace, and a multistage electric furnace.

SAFETY PRECAUTIONS: Make sure that the power is turned off before removing panels on the electric furnace. Turn the power on only after your instructor approves your setup.

PROCEDURES

1. With the power off on a single-stage electric heating unit, remove the panels. Remove the red wire and install a "10-wrap loop" around one jaw of a clamp-on ammeter. This is a loop of thermostat wire wrapped 10 times around the ammeter jaw, which will amplify the current reading by 10 times. Connect back into the circuit. See text Figure 13–34.

2. Turn the power on. Switch the thermostat to heat and adjust it to call for heat. Check the current at the loop and record. _____ A

3. This should be approximately 5 amperes. Divide the reading by 10 to get the actual current. Record it here. _____ A

4. Adjust the heat anticipator under the thermostat cover to reflect the actual current if it is not set correctly.

5. With the power off on a multistage electric heating unit, remove the panels. Remove the red wire and install a "10-wrap loop" around the jaw of an ammeter as in Step 1. Connect back into the circuit.

6. Turn the power on. Switch the thermostat to heat and adjust to call for heat. Check the current and record. _____ A

7. Divide this reading by 10 to get the actual current and record. _____ A

8. Adjust the heat anticipator to reflect the actual current if it is not set correctly.

MAINTENANCE OF WORK STATION AND TOOLS: Return all panels. Replace all tools and instruments, leaving your work area neat and orderly.

SUMMARY STATEMENTS: Write a description of how the heat anticipator works, telling what its function is. Describe why the current is greater when more than one sequencer is in the circuit.

QUESTIONS

1. Most heat anticipators are _____. (fixed resistance or variable resistance)

2. A heat anticipator that is set wrong can cause _____. (No heat, excessive temperature swings or noise in the system)

3. If the service technician momentarily shorted across the sequencer heater coil, it would cause _____. (Blown fuse, burned heat anticipator, the heat turned on or the cooling turned on)

4. All heating thermostats have heat anticipators. _____ (True or false)

5. When a system has several sequencers that would cause too much current draw through the heat anticipator, which of the following could be done to prevent the anticipator from burning up?
   A. Install a higher amperage rated thermostat
   B. Wire the sequencer so that all of the coils did not go through the anticipator
   C. Change to a package sequencer
   D. Let it get hot

6. The heat anticipator is important to proper temperature control. (True or false)

7. When the furnace is two stage, there are how many heat anticipators?

8. The heat anticipator is located in _____. (The thermostat or the subbase)

LAB 30

Low-Voltage Control Circuits Used in Electric Heat

Name _____ Date _____ Grade _____

OBJECTIVES: Upon completion of this exercise, you should be able to describe and draw diagrams illustrating the low-voltage control circuits in single- and multistage electric furnaces.

INTRODUCTION: You will study the low-voltage wiring in a single- and multistage electric furnace. You will then draw a pictorial and ladder diagram for each of these circuits.

TEXT REFERENCES: Units 13 and 14

TOOL AND MATERIALS: Phillips and straight blade screwdrivers, and 1/4" and 5/16" nut drivers.

SAFETY PRECAUTIONS: Make sure that the power is turned off and locked out before removing panels on the electric furnaces. Turn the power on only after your instructor approves your setup.

PROCEDURES

1. Make sure that the power is off and locked out, then remove the panels from an electric heater that has one stage of heat. The electric circuits should be exposed.

2. Study the wiring in the low-voltage circuit until you understand how it controls the heat strip.

3. Draw a pictorial diagram including the thermostat of the low-voltage circuit in the space below. Draw a ladder diagram of the circuit.

                PICTORIAL   LADDER

4. Make sure that the power is off and remove the panels from an electric furnace that has more than one stage of heat. The electric circuits should be exposed.

5. Study the wiring in the low-voltage circuit until you understand how it controls the strip heaters.

6. Draw a pictorial diagram including the thermostat of the low-voltage circuit in the space below. Draw a ladder diagram of the same circuit.

PICTORIAL   LADDER

MAINTENANCE OF WORK STATION AND TOOLS: Replace all panels. Replace tools and leave work station as instructed.

SUMMARY STATEMENTS: Describe the sequence of events in the complete low-voltage circuit for each system studied.

QUESTIONS

1. Low-voltage controls are used for which of the following reasons?
   A. Less expensive
   B. Easier to troubleshoot
   C. More reliable
   D. Safety

2. Low-voltage wire has an insulation value for which of the following?
   A. 50 V
   B. 100 V
   C. 500 V
   D. 24 V

3. Wire size is determined by which of the following?
   A. Voltage
   B. Current
   C. Resistance
   D. Power

4. The gage of wire that is normally used in residential low-voltage control circuits is which of the following?
   A. No. 12
   B. No. 14
   C. No. 18
   D. No. 22

5. Low-voltage wire always has to be run in conduit. _____ (True or false)

6. The nominal low-voltage for control circuits is which of the following?
   A. 50 V
   B. 24 V
   C. 115 V
   D. 230 V

7. All units have only one low-voltage power supply. _____ (True or false)

LAB 31 — Checking a Package Sequencer

Name _____ Date _____ Grade _____

OBJECTIVES: Upon completion of this exercise, you should be able to check a package sequencer by energizing the operating coil and checking the contacts of the sequencer with an ohmmeter.

INTRODUCTION: You will energize the operating coil of a package sequencer for electric heat with a low-voltage power source to cause the contacts to make. You will then check the contacts with an ohmmeter to make sure that they close as they are designed to do.

TEXT REFERENCES: Units 13 and 14

TOOLS AND MATERIALS: Low-voltage power supply, a VOM, and a package sequencer used in an electric furnace.

SAFETY PRECAUTIONS: You will be working with low voltage (24 V). Do not allow the power supply leads to arc together.

PROCEDURES

1. Turn the VOM function switch to AC voltage and the range selector switch to the 50 V or higher scale.

2. With the power off, fasten one VOM lead to one of the 24-V power supply leads and the other VOM lead to the other supply lead.

3. Turn the power on and record the voltage. _____ V. It should be very close to 24 V.

4. Turn the power off.

5. Fasten the 2 leads from the 24-V power supply to the terminals on the sequencer that energize the contacts. They are probably labeled heater terminals.

6. Turn the ohms selector range switch to R x 1. Fasten the VOM leads on the terminals for one of the electric heater contacts. See text Figure 13-17 for an example of how a sequencer is wired.

7. Turn the 24-V power supply on. You should hear a faint audible "click" each time a set of contacts makes or breaks.

8. After about 3 minutes, check each set of contacts with the VOM to see that all sets are made. You may want to disconnect the power supply and allow all contacts to open and repeat the test to determine the sequence in which the contacts close.

9. Turn off the power and disconnect the sequencer.

MAINTENANCE OF WORK STATION AND TOOLS: Return the VOM to its proper storage place. You may need the sequencer while preparing the Summary. Put it away in its proper place when you are through with it.

SUMMARY: Draw a wiring diagram of the package sequencer, showing all components. See text Figure 13–17, for an example.

QUESTIONS

1. What type of device is inside the sequencer to cause the contacts to close when voltage is applied to the coil?

2. What type of coil is inside the sequencer? (magnetic or resistance)

3. What is the advantage of a sequencer over a contactor?

4. How many heaters can a typical package sequencer energize?

5. How is the fan motor started with a typical package sequencer?

6. What is the current-carrying capacity of each contact on the sequencer you worked with?

7. Do all electric furnaces use package sequencers?

8. If the number 2 contact failed to close on a package sequencer, would the number 3 contact close?

9. How should you determine the heat anticipator setting on the room thermostat when using a package sequencer?

10. How should you determine the heat anticipator setting on the room thermostat when using individual sequencers?

LAB 32

Checking Electric Heating Elements Using an Ammeter

Name _____ Date _____ Grade _____

OBJECTIVES: Upon completion of this exercise, you should be able to check an electric heating system by using a clamp-on ammeter to make sure that all heating elements are drawing power.

INTRODUCTION: You will identify all electric heater elements, then start an electric heating system. You will then use an ammeter to prove that each element is drawing power.

TEXT REFERENCES: Units 13 and 14

TOOL AND MATERIALS: A VOM, a clamp-on ammeter, goggles, straight blade and Phillips screwdrivers, a flashlight, 1/4" and 5/16" nut drivers, and an electric heating system.

SAFETY PRECAUTIONS: You will be placing a clamp-on ammeter around the conductors going to the electric heat elements. Do not pull on the wires or they may pull loose at the terminals.

PROCEDURES

1. Turn off the power to an electric furnace and remove the door to the control compartment.

2. Locate the conductor going to each electric heat element. See text Figures 13-15 and 13-17. Consult the diagram for the furnace you are working with. Make sure that you can clamp an ammeter around each conductor.

3. Turn the power on and listen for the sequencers to start each element. Wait for about 3 minutes; the fan should have started.

4. Clamp the ammeter on one conductor to each heating element and record the readings below. When the element is drawing current, it is heating.

    - Element number 1 _____ A   Element number 2 _____ A
    - Element number 3 _____ A   Element number 4 _____ A
    - Element number 5 _____ A   Element number 6 _____ A

    NOTE: Your furnace may only have 2 or 3 elements.

5. Turn the room thermostat off and, using the clamp-on ammeter, observe which elements are de-energized first.

6. Replace all panels with the correct fasteners.

MAINTENANCE OF WORK STATION AND TOOLS: Return all tools to their proper places. Replace all panels, if appropriate, on the electric furnace.

SUMMARY STATEMENT: Describe the function of an electric heating element, telling how the current relates to the resistance in the actual heating element wire.

QUESTIONS

1. What are the actual heating element wires made of?

2. If the resistance in a heater wire is reduced, what will the current do if the voltage is a constant?

3. How much heat would a 5 kW heater produce in Btu per hour?

4. What started the fan on the unit you were working on?

5. What is the difference between a contactor and a sequencer?

6. Why are sequencers preferred over contactors in duct work applications?

7. How much current would a 5 kW heater draw with 230 volts as the applied voltage?

8. What is the advantage of a package sequencer over a system with individual sequencers?

9. If the number one sequencer coil were to fail in a system with individual sequencers, would all the heat be off?

LAB 33   Changing a Sequencer

Name _____ Date _____ Grade _____

OBJECTIVES: Upon completion of this exercise, you should be able to change a sequencer in an electric furnace.

INTRODUCTION: You will remove a sequencer in an electric furnace and replace and rewire it just as though you have a new one.

TEXT REFERENCES: Unit 13

TOOLS AND MATERIALS: Flat blade and Phillips screwdrivers, 1/4" and 5/16" nut drivers, VOM, and ammeter and needle nose pliers.

SAFETY PRECAUTIONS: Make sure the power is off and locked out before you start this project. Use the voltmeter to verify the power is off.

PROCEDURES

1. Turn the power off and lock it out.
2. Check to make sure the power is off using the VOM.
3. Draw a wiring diagram showing all wires on the sequencer to be changed. If some of the wires are the same color, make a tag and mark them for easy replacement.

4. Remove the wires from the sequencer one at a time. Use care when removing wires from spade or push on terminals and don't pull the wire out of the connector. Use needle nose pliers. Check for properly crimped terminals.
5. Remove the sequencer to the outside of the unit. This would be just like having a new one.
6. Now, remount the sequencer.
7. Replace the wiring one at a time. Be very careful that each connection is tight. Electric heat pulls a lot of current and any loose connection will cause a problem.
8. When the sequencer is installed to your satisfaction, ask the instructor to approve the installation before you turn the power on.
9. Turn the power on and set the thermostat to call for heat.
10. After the unit has been on long enough to allow the sequencers to energize the heaters, use the ammeter to verify the heaters are operating.
11. Turn the unit off when you are satisfied it is working.

MAINTENANCE OF WORK STATION AND TOOLS: Return all tools to their places and leave the unit like your instructor directs you.

SUMMARY STATEMENT: Describe the internal operation of a sequencer.

QUESTIONS

1. What would low voltage do to the capacity of an electric heater?

2. What would high voltage do to the capacity of an electric heater?

3. An electric heater has a resistance of 11.5 ohms and is operating at 230 volts. What would the current reading be for this heater? Show your work.

4. What would the power be for the above heater? Show your work.

5. What would the Btuh rating be on the above heater? Show your work.

6. If the voltage for the above heater were to be increased to 245 volts, what would the power and Btuh be? Show your work.

7. If the voltage were reduced to 208 volts, what would the power and Btuh be for the above unit? Show your work.

8. Using the above voltages, would the heater be operating within the + or - 10% recommended voltage?

9. What is the unit of measure the power company charges for electricity?

10. What starts the fan motor in the unit you worked on?

LAB 34

Changing the Heating Element in an Electric Furnace

Name _____ Date _____ Grade _____

OBJECTIVES: Upon completion of this exercise, you should be able to change a heating element in an electric furnace using the correct techniques.

INTRODUCTION: You will change the heating element in an electric furnace using the correct tools and equipment. You will then restart the furnace and check it for proper operation.

TEXT REFERENCES: Unit 13

TOOLS AND MATERIALS: Flat blade and Phillips screwdrivers, 1/4" and 5/16" nut drivers, needle nose pliers, VOM, and an ammeter

SAFETY PRECAUTIONS: Make sure all power is off and locked out before starting this exercise. Use care when disconnecting all connections making sure they are labeled and not abused. No loose connections can be allowed.

PROCEDURES

1. TURN THE POWER OFF AND LOCK IT OUT, THEN CHECK WITH VOM TO BE SURE.
2. Remove the panel door to the electric heat section.
3. Check all wires to the heater to be changed for color coding, comparing them to the wiring diagram. If they are not coded, tag them so the correct wires can be placed back in their places. This is very important. You can't rely on memory.
4. Remove the wires, one at a time. If spade terminals are used, take care removing the terminals. Needle nose pliers may be used by gripping only the connector. Do not pull on the wire.
5. When all the wires are clear, loosen the heater mounting and remove the heater. If there is any question about how the heater is removed, consult your instructor.
6. Set the heater section on a work bench and record the following information.
   A. What is the resistance through the heating element? _____ ohms.
   B. Does this element have a fuse link? (yes or no)
7. Now, replace the heating element in the furnace and fasten it.
8. Replace all wiring to the heating element. Make sure all connections are tight.
9. Have your instructor inspect the job.
10. When all is corrected, turn on the power and start the unit. Give it a few minutes to warm up and for all heaters to be energized.
11. Record the current and voltage of the heater.
    A. _____ A   B. _____ V   Compare with data on unit data plate.
12. Turn the unit off.

MAINTENANCE OF WORK STATION AND TOOLS: Return all tools to their places and leave the unit as your instructor directs you.

SUMMARY STATEMENT: Describe the complete cycle of a forced air electric furnace, including the low voltage control circuit.

QUESTIONS

1. What is the heating element wire made of?

2. What is the purpose of the fuse link?

3. What other protection from overheating did this unit have?

4. How many elements did this furnace have?

5. If more than one, did it have stack sequencers, individual sequencers or a package sequencer?

6. Using Ohm's Law, what should the current for this heater have been? Show your work.

7. Are the actual current and the calculated current the same? If not, why not?

8. What is the Btuh for this heater? Show your work.

LAB 35

Familiarization of a Hot Water Boiler and Heating System

Name _____ Date _____ Grade _____

OBJECTIVES: Upon completion of this exercise, you should be able to recognize the various components of a typical boiler and hot water heating system.

INTRODUCTION: You will examine the various components of a hot water boiler and heating system for the purpose of identifying the various components.

TEXT REFERENCES: Units 3, 9, 12 and 15

TOOLS AND MATERIALS: Phillips and flat blade screwdrivers, 1/4" and 5/16" nut drivers and a flashlight.

SAFETY PRECAUTIONS: Use care while working around any hot water equipment as it is hot enough to create burns.

PROCEDURES

Enter the following data about a hot water heating system.

- Type of boiler, gas, oil or electric _____
- Manufacturer _____ Model Number _____
- Boiler input _____ Output _____
- If a gas boiler, fuel consumption per hour _____ CFH
- If an oil boiler, fuel consumption per hour _____ GPH
- If electric boiler, power consumption _____ KWH
- Size of water pipe leaving the boiler _____ inches
- Size of water pipe entering the boiler _____ inches
- At what pressure is the relief valve suppose to relieve? _____ psig, What temperature? _____ °F
- Does the relief valve discharge to a drain or outside? _____
- Does the water pump discharge into the boiler or away from the boiler? _____
- Describe the location of the expansion tank in relation to the boiler.

- What type of terminal units does this system have to dissipate the heat to the conditioned space?
  Baseboard _____ Panel heat _____ Unit heater _____
  Radiator _____ Other _____
- Are there pressure gages at the pump inlet and outlet for checking pressure drop? _____
- Is there a flow measuring device for determining the water flow for the entire system? _____ for the individual terminal units or circuits? _____
- What is the working pressure for the system? _____ psig
- What is the typical temperature for the water in this system? _____ °F

MAINTENANCE OF WORK STATION AND TOOLS: Replace any panels that had to be removed and restore the work place for the next student.

SUMMARY STATEMENT: Describe how city water is made up to the system you looked at if water should leak from the system.

QUESTIONS

1. Why does the relief valve have both a pressure and temperature setting?

2. What is the purpose of the expansion tank on a hot water heating system?

3. What would the symptoms be if air were to be circulating with the water in the hot water system?

4. Name 2 different types of air bleed systems for hot water circulating systems.

5. Where does air come from in a hot water heating system?

6. What kind of harm can air cause to a hot water heating system?

7. Describe the difference in a forced convection and a natural convection terminal unit.

8. Which system more nearly balances the water flow, a reverse return or a direct return piping system?

9. What makes the water flow to the circuit through the terminal unit in a one pipe tee system?

10. Should balancing valves be installed in all complex hot water heating systems?

LAB 36

Checking Water Flow Using an Orifice Flow Checking Device

Name _____ Date _____ Grade _____

OBJECTIVES: Upon completion of this exercise, you should be able to use an orifice metering device to check the water flow rate in a hot water system.

INTRODUCTION: You will use a pressure drop device to determine the water flow rate in a hot water heating system or terminal unit.

TEXT REFERENCES: Unit 15

TOOLS AND MATERIALS: Differential pressure gage used for checking the pressure drop through a flow checking device.

SAFETY PRECAUTIONS: Be careful while working around hot water, it can cause burns.

PROCEDURES

1. Select a flow checking device in a hot water circuit and look over the pressure drop chart for this device. It may be a chart or a tag that is permanently attached to the device.

2. Fasten the differential pressure gage to the correct connections on the flow checking device. You may need the help of your instructor for this.

3. Bleed the air from the hoses following the gage manufacturer's instructions.

4. With the system pump running and water flow established, open the correct valves at the instrument to obtain the pressure drop reading.

5. Record the pressure drop reading here. _____ Inches of water column

6. Looking at the graph for the differential gage, record the water flow here. _____ gallons per minute

7. If there are more test points that your instructor wishes you to test, the results may be recorded on another piece of paper.

8. Remove the gages and replace any test point caps to prevent leaks.

9. Clean the instrument of any water that may have leaked into the gage case or directions, leaving the gage and case clean and dry.

MAINTENANCE OF WORK STATION AND TOOLS: Return the work station to a clean and orderly place. Wipe up any water that may have spilled.

SUMMARY STATEMENT: Describe how the flow checking device works and how it should be installed to obtain the correct reading.

QUESTIONS

1. Why is a typical differential gage used for flow checking graduated in inches of water column instead of psig?

2. Why should you bleed the air from the instrument before taking a reading?

3. How is the water flow adjusted if there is too much water flow?

4. Why should the flow checking device be piped into the circuit with some distance from the nearest fittings?

5. How far should the flow checking device be from the nearest fittings to obtain a correct fitting?

6. What could the problem be if there is not enough flow in a system or circuit in a system?

LAB 37

Checking the System Flow Rate Using the Pressure Drop Across the Pump

Name _____ Date _____ Grade _____

OBJECTIVES: Upon completion of this exercise, you should be able to use a pump curve and pressure differential gages to determine the flow rate through a water pump in GPM.

INTRODUCTION: You will fasten gages to a working pump system and take pressure readings on the pump inlet and outlet to determine the pressure difference between the pump inlet and outlet.

TEXT REFERENCES: Unit 15, Figure 15-19

TOOLS AND MATERIALS: Pressure gage or gages, clamp on ammeter slip joint pliers, two 6" adjustable wrenches and a pump curve for the pump to be checked.

SAFETY PRECAUTIONS: Use caution when working around hot water under pressure.

PROCEDURES

1. Fasten the gage connections to the inlet and outlet to the circulating pump. It is preferred that 1 gage be used for both readings, see Figure 15-19 for the arrangement.

2. Turn on the control valves to the gage and record the pump inlet reading. _____ psig

3. Record the pump outlet reading. _____ psig

4. Pump inlet pressure − pump outlet pressure = difference
   _____ outlet psig − _____ inlet psig + _____ psig diff

5. Convert the pressure difference from psig to feet of head. Psig × 2.31 = feet of head
   _____ psig × _____ feet of head = _____ feet of head

6. Use the difference in pressure and a pump curve for the pump to determine the water flow. NOTE, if a pump curve is not available, you may call the supplier for this type pump and get this information. They may send you a set of curves to have on hand.

7. Full load current from the nameplate _____ FLA

   Actual current of the motor _____ amperes

8. If there is a flow checking device for the entire system, you may check the pump data against the flow checking device to determine if they are comparable. The most accurate method would be the flow checking device if it is properly installed.

9. Shut off the valves and remove the gage arrangement.

10. Wipe up any water from the floor or piping. Clean the gages and tools if needed. Do not leave any water on them as this may cause damage.

MAINTENANCE OF WORK STATION AND TOOLS: Return all tools and equipment to their respective places.

SUMMARY STATEMENT: Describe how a centrifugal water pump moves water from the inlet to the outlet.

QUESTIONS

1. Why is it preferable to use one gage instead of two when checking pressure drop across a pump?

2. The difference from the inlet to the outlet of a pump is 22 psig, how many feet of head is this?

3. When throttling the flow of water at the hot water pump, which valve should be used, the inlet or the outlet?

4. Explain the reason for your choice in question 3.

5. When the water is throttled at the pump outlet, what does the motor current do, rise or fall?

6. When the water is throttled at the pump inlet, what does the motor current do, rise or fall?

7. What would the result be if a water pump consistently pumped air in the water circuit?

8. If a hot water pump were in the basement of a building and the highest hot water coil above the pump were to be 65 feet, what would the standing pressure on the pump be?

9. What would the results be if a water pump were to be valved completely off and allowed to run for a long period of time?

10. A hot water coil has a water temperature difference of 40°F from the inlet to the outlet and according to the circuit checking device, it is moving 40 gpm of water, how many Btu of heat is this coil putting out?

LAB 38 Cleaning a Strainer

Name _____ Date _____ Grade _____

OBJECTIVES: Upon completion of this exercise, you should be able to remove and clean an inline strainer in a hot water system.

INTRODUCTION: You will close the appropriate valves and remove an in line strainer, clean it if needed and re-install it, then start the system back up.

TEXT REFERENCES: Unit 15

TOOLS AND MATERIALS: Two pipe wrenches, a drain pan to catch excess water if the strainer is not over a floor drain, pipe seal and a wire brush.

SAFETY PRECAUTIONS: Make sure you shut off water to both sides of the strainer and remove it with care, hot water can burn you.

PROCEDURES

1. Locate the hot water strainer that is to be removed, you may need to ask your instructor.

2. Locate the valves that will isolate the strainer. NOTE: These valves may also isolate the pump and it may be drained down at the same time.

3. Check and record the pressure difference across the hot water pump, then shut it off. _____ psig

4. Remove a gage that is common to the strainer circuit and crack the valve to relieve the pressure that may be in the pipe.

5. Slowly remove the plug from the end of the strainer and catch any water that may drain if there is no floor drain under the strainer.

6. Now remove the strainer from the strainer housing and examine it. If it is dirty, use the wire brush to clean it. When it is clean, replace it back in the strainer housing.

7. Apply thread seal to the threads for the strainer and replace the strainer plug.

8. Make sure the valve to the gage port is closed and slowly open the valves that were used to isolate the strainer. Bleed air from the gage port line and from the top of the pump housing to minimize the amount of air that may be left in the system.

9. Start the pump and listen for water flow, check for pressure drop across the pump. It should be the same as before you started.

10. Bleed air from a high point in the system to remove any air from the system that may have been left in the pump or strainer housing.

11. Clean up any water in the floor and on the pump.

MAINTENANCE OF WORK STATION AND TOOLS: Clean all water from the tools and replace them in the proper place.

SUMMARY STATEMENT: Describe how a strainer may be checked from the outside for a restriction.

QUESTIONS

1. What is the purpose of a strainer in a hot water system?

2. What material was the strainer made of in the system you worked on?

3. Did the strainer contain any trash? If so, describe what it may have consisted of.

4. Why is the strainer in the pump inlet rather than the pump outlet?

5. What would be the likely results of a restricted strainer on a hot water system?

6. What was the working pressure for the system you were working on? _____ psig

7. What is the typical working pressure for a low pressure water system?

8. Was there any water treatment chemicals in the water system you worked on?

9. What is the purpose of water treatment in a hot water system?

10. What would be the results of a system that had continuous water leaks and large amounts of city water entered the system each year?

LAB 39　Filling a Hot Water Heating System

Name _____ Date _____ Grade _____

OBJECTIVES: Upon completion of this exercise, you will be able to fill an empty hot-water system with fresh water and put it back in operation.

INTRODUCTION: You will start with an empty boiler and hot-water system and fill it with water, bleed the air and put it back into operation.

TEXT REFERENCES: Unit 15

TOOLS AND MATERIALS: Two pipe wrenches (12"), a straight blade and Phillips screwdrivers, a water hose if needed, and a boiler and hot-water heating system.

SAFETY PRECAUTIONS: NEVER ALLOW COLD WATER TO ENTER A HOT DRY BOILER. IF YOUR BOILER IS HOT, SHUT IT OFF THE DAY BEFORE THIS EXERCISE OR DO NOT DRAIN THE BOILER AS A PART OF THE EXERCISE.

PROCEDURES

1. Turn the power off to the boiler and water pumps and drain the entire system at the boiler drain valve if the boiler is at the bottom of the circuit. If the boiler is at the top of the system, drain the system valve at the lowest point.

2. Make sure the boiler is cool to the touch of your hand.

3. Close all valves in the system and start to fill the water system. CHECK TO MAKE SURE THE SYSTEM HAS A WATER REGULATING VALVE, such as in text Figure 15–121. With this valve in the system, you cannot apply too much pressure to the system. IF THE SYSTEM DOES NOT HAVE AN AUTOMATIC FILL VALVE, WATCH THE PRESSURE GAGE AND DO NOT ALLOW THE PRESSURE TO RISE ABOVE THE RECOMMENDED, PROBABLY 12 PSIG.

4. Fill the system until the automatic valve stops feeding or the correct pressure is reached (in which case, shut the water off).

5. Go to the air bleed valves and make sure that they allow any air to escape. See text Figures 15-108 and 109, for examples of manual and automatic types. If they are manual valves, bleed them until a small amount of water is bleeding out.

6. Start the pump and check the air bleed valves again.

7. Start the boiler and wait until it gets hot to the touch.

8. Carefully feel the water line leaving the boiler. It should get hot and the hot water should start leaving the boiler and move towards the system. The pump should be moving the water in a forced-water system and gravity will move it in a gravity system. You should now be able to follow the hot water throughout the system and with the cooler water going back to the boiler.

9. You should now have a system that is operating. The water level in the expansion tank has not been checked for a system with an air bleed system at the expansion tank. See the next exercise.

10. Clean up the work area, leave no water on the floor. LEAVE THE BOILER AS YOUR INSTRUCTOR INDICATES.

MAINTENANCE OF WORK STATION AND TOOLS: Return all tools to their respective places.

SUMMARY STATEMENTS: From the system that you were working with, describe the distribution piping system.

QUESTIONS

1. How does air get into a water system?

2. How can you get air out of a water system?

3. What are the symptoms of air in a water system?

4. Why should cold water not be allowed to run into a hot dry boiler?

5. What is inside an automatic air bleed valve to allow the air to get out?

6. What type of pump is typically used on small hot-water heating systems?

7. What is the purpose of balancing valves in a hot-water system?

8. How is air separated from the water at the boiler in some systems?

9. When it is separated, as in Question 8, where does it go?

10. What is the purpose of a system pressure-relief valve?

LAB 40

Filling an Expansion Tank on a Hot-Water Heating System

Name _____ Date _____ Grade _____

OBJECTIVES: Upon completion of this exercise, you will be able to fill an expansion tank with a bleed system to the correct level.

INTRODUCTION: You will shut off the water supply to the system and check the water level of the tank. You will then let the water level down below the bleed valve level and fill it back to the bleed level.

TEXT REFERENCES: Unit 15

TOOLS AND MATERIALS: Two adjustable wrenches (8"), two pipe wrenches (10"), a pan to hold hot water, straight blade and Phillips screwdrivers, and a hot-water heating system.

SAFETY PRECAUTIONS: You will be bleeding hot water out of the system; do not allow any of it to get on you.

PROCEDURES

1. Start the hot-water heating system and allow the water to get up to temperature.
2. Make sure the pump is running.
3. Study Figures 15-111, 15-112 and 15-113 in the text for an example of a tank and the fitting arrangement. Compare it to the tank and fitting arrangement on your system.
4. Place the pan under the tank fittings. Study Figure 15-118, a vent valve, in the text. Slowly open the valve and see if water or air vents out.
5. If air vents out, allow the valve to vent until water begins to bleed, then shut the valve off. The water level in the tank is at the bleed valve inlet and correct. If water vents out, shut the valve off. You need to lower the water level in the tank.
6. Shut the water supply to the boiler off.
7. Open the valve at the bottom of the system and allow several gallons of water to drain out. This will lower the water level in the tank.
8. Now open the feeder water valve and allow water from the supply line to enter the system until it stops.
9. Now check the water level at the expansion tank again by opening the vent. If you still have water bleeding out, repeat steps 6, 7 and 8. This must be done until air bleeds as in step 5 and the correct water level is reached.

MAINTENANCE OF WORK STATION AND TOOLS: Return all tools to their proper places. Mop up any water on the floor.

SUMMARY STATEMENT: Describe the process of separating water at the boiler and moving it to the expansion tank.

QUESTIONS

1. How does water enter a hot-water system?

2. Why is an expansion tank with air in it necessary?

3. How does the tank fitting in text Figure 15–118 assure the correct water level when adjusted?

4. Why must the expansion tank always be located higher than the boiler?

5. Why must the air bleed system be the high point in the system?

6. Why is water treatment necessary to maintain a correct hot-water system?

7. Where does water makeup come from when the system loses water?

8. How does water leak from a system?

9. Why should you NEVER allow cold water to enter a hot empty boiler?

10. Describe a tankless domestic water heater.

LAB 41 — Installing a Pump Coupling

Name _____ Date _____ Grade _____

OBJECTIVES: Upon completion of this exercise, you should be able to remove a pump coupling and install a new coupling.

INTRODUCTION: You will remove a flexible pump coupling from an operating pump and re-install it as though it is a new coupling.

TEXT REFERENCES: Unit 15

TOOLS AND MATERIALS: Small set of socket wrenches, set of end wrenches, assortment of Allen head wrenches, a soft face hammer, some oil, a small amount of grease, a Phillips and a flat blade screwdriver.

SAFETY PRECAUTIONS: Shut off the power and lock it out.

PROCEDURES

1. Shut off the power to the pump and lock it out. Check it with a volt meter to be sure it is off. Where there are multiple pumps, make sure you have shut off the correct pump.

2. Remove the cover to the pump coupling and set it aside.

3. Loosen the coupling from the motor end of the shaft and slide it back far enough to take the flexible portion of the coupling apart in the middle. If it will not separate enough, loosen the hub on the pump shaft and slide it back.

    NOTE: Do not use a hammer to move the couplings on the shaft. If they are tight, clean the shaft with very fine sand tape (400 grit or finer) and use a soft face hammer to move the coupling on the shaft. Lubricate with oil after removing any signs of grit from sanding. Be sure you do not damage the shaft. If you have trouble, ask your instructor for help.

4. Remove the pump portion of the coupling, then remove the motor portion of the coupling.

5. Assemble the coupling on the workbench so that you can see how the coupling functions. This would represent a new coupling out of the box.

6. If the pump or motor shaft needs cleaning to remove rust, do it now. When the shafts are clean, apply a small amount of grease to each shaft.

7. Replace the motor portion of the coupling, leaving it loose on the shaft.

8. Replace the pump portion of the coupling.

9. Replace the flexible portion of the coupling.

10. Tighten both the pump and motor portions of the coupling to their respective shafts.

    NOTE: Make sure the set screw is tightened to the flat on the shaft if there is one.

11. Replace the shield to the coupling.

12. Get your instructor to inspect the job and then start it up with permission.

MAINTENANCE OF WORK STATION AND TOOLS: Return all tools to the correct storage areas.

SUMMARY STATEMENT: Describe how the flexible coupling on the pump works.

QUESTIONS

1. Why is it important to lock out the power while working on a pump coupling?

2. What is the purpose of the shield for the pump coupling?

3. What was the motor speed in RPM for the pump you worked on?

4. Can a pump coupling compensate for any misalignment between a pump and motor?

5. What type of fastener was used to fasten the coupling to the motor shaft?

6. What type of fastener was used to fasten the coupling to the pump shaft?

7. Were the motor and pump shafts the same size?

8. How would a technician tell if the motor or the pump were tight if the coupling could not be turned over by hand?

LAB 42

Familiarization of a Steam Boiler and Heating System

Name _____ Date _____ Grade _____

OBJECTIVES: Upon completion of this exercise, you should be able to recognize the components of a steam boiler and heating system.

INTRODUCTION: You will examine the various component of a steam boiler heating system for the purpose of identifying the various components.

TEXT REFERENCES: Unit 16

TOOLS AND MATERIALS: Phillips and flat blade screwdrivers, 1/4" and 5/16" nut drivers and a flashlight.

SAFETY PRECAUTIONS: Be very careful around all steam components if the system is on. Steam systems are very hot and can cause burns.

PROCEDURES

Enter the following data about a steam heating system.

- Type of boiler, gas, oil or electric _____
- Manufacturer _____ Model Number _____
- Boiler input _____ Output _____
- If a gas boiler, fuel consumption per hour _____ CFH
- If an oil boiler, fuel consumption per hour _____ GPH
- If electric boiler, power consumption _____ KWH
- Size of water pipe leaving the boiler _____ inches
- Size of water pipe entering the boiler _____ inches
- At what pressure is the relief valve suppose to relieve? _____ psig.
- Does the relief valve discharge to a drain or outside? _____
- What type of terminal units does this system have to dissipate the heat to the conditioned space?
  Baseboard _____ Panel heat _____ Unit heater _____
  Radiator _____ Other _____
- What is the working pressure for the system? _____ psig
- Where is the low water cut-off located on the boiler?

- What type of fresh water make-up does this system use?

- What type of condensate return does this system have, gravity or condensate pump? _____
- What type of steam traps does the terminal system use? (float, thermostatic or float)
- Is the boiler above or below the highest terminal heating unit? _____

2. Replace all panels with the correct fasteners.

MAINTENANCE OF WORK STATION AND TOOLS: Return all tools to their proper places.

SUMMARY STATEMENT: Describe how the water level is maintained in the boiler when there is a loss of water.

QUESTIONS

1. Did the boiler you worked with use water treatment?

2. Why is it important that the water level not drop below the prescribed level?

3. What type of control stopped and started the boiler you worked on?

4. What is the purpose of the condensate trap at the terminal unit in a steam system?

5. Describe water hammer.

6. Why do many steam systems have an expansion device?

7. Describe how a 1 pipe steam system works.

8. What was the operating (different from working pressure) pressure of the boiler you worked with? _____ psig

9. What is the advantage of steam heat over hot water heat?

10. Describe the steam dome in a hot water boiler.

11. Describe the differences in a steam and hot water boiler.

LAB 43

Checking a Condensate Trap for Proper Operation

Name _____ Date _____ Grade _____

OBJECTIVES Upon completion of this exercise, you should be able to examine an operating steam trap and check it for proper operation.

INTRODUCTION: You will use a temperature tester or other testing device to check the temperature of an operating steam trap to determine that the trap is functioning to trap condensate in the correct manner.

TEXT REFERENCES: Unit 16

TOOLS AND MATERIALS: Temperature tester or other method of checking the inlet and outlet temperature of a condensate trap in the appropriate temperature range and a screwdriver or other device that may be used for listening as in Figure 16-47 in the text.

SAFETY PRECAUTIONS: Use care when working around steam components as they are very hot and can cause burns.

PROCEDURES

1. Proceed to an operating steam terminal unit and locate the condensate trap.

2. Fasten a high temperature thermometer lead to the inlet steam line. NOTE: Your instructor may want you to check the trap using some other method but it will amount to the same thing.

3. Fasten a high temperature thermometer lead to the line leaving the condensate trap and returning to the boiler.

4. Record the entering temperature. _____ °F

5. Record the leaving temperature. _____ °F

6. Record the temperature difference. _____ °F

7. Use a screwdriver to listen to the trap. Does it sound like it is passing water or steam vapor? _____ See Figure 16-47 in the text.

8. Describe how the condensate is routed from the trap back to the boiler.

9. Remove the thermometer leads and restore the system to the condition you found it.

MAINTENANCE OF WORK STATION AND TOOLS: Return all tools to their respective places.

SUMMARY STATEMENT: Describe the function of a condensate trap.

QUESTIONS

1. What would the results be if a condensate trap were to be stuck closed where no condensate would return to the boiler?

2. What would the results be if a condensate trap were to be stuck in the open position?

3. How does air pass through a condensate trap?

4. How does air get in a steam system?

5. Describe how condensate is routed around an obstacle when being returned to the boiler.

6. Describe how steam is routed around an obstacle when being returned to the boiler.

7. Did the trap in the system you worked with have a strainer before the trap?

8. What may be captured in a strainer and where would it come from?

LAB 44

Performing Blowdown on a Steam Boiler

Name _____ Date _____ Grade _____

OBJECTIVES: Upon completion of this exercise, you should be able to perform a blowdown on a steam boiler for the purpose of reducing the scum and residual trash in the low water cut-off control.

INTRODUCTION: You will use an operating steam system to perform a blowdown procedure.

TEXT REFERENCES: Unit 16

TOOLS AND MATERIALS: None required.

SAFETY PRECAUTIONS: Make sure the by-products of the blowdown are directed away from your feet or body as raw steam and hot water will be released and burns can occur.

PROCEDURES

1. Familiarize yourself with the components of a steam boiler system.

2. Locate the low water cut-off component.

3. Locate the blow-down valve. It should be on the bottom of the low water cut-off control. Make sure you understand which direction the blow-down by-products will be discharged.

4. Shut the boiler off. If it is an electric boiler, you can proceed immediately. If gas or oil, allow the flame to go completely out.

5. With your instructor observing, open the blow-down valve for approximately 2 seconds and observe the contents of the discharge.

6. Let the boiler water settle down for about 30 seconds and blow it down again.

7. If the water was clear of trash and foam, proceed with step 8. If not, blow it down again after about 30 seconds.

8. Restart the boiler and wait for it to come up to full operation before leaving it.

MAINTENANCE OF WORK STATION AND TOOLS: Make sure there is no water left in the floor for anyone to slip on.

SUMMARY STATEMENT: Describe the purpose of blow-down and what you saw from the blow-down by-products.

QUESTIONS

1. Why do contaminates collect in the low water cut-off control area?

2. Did this boiler have water treatment in the water?

3. What are some of the contaminants that water treatment control?

4. What is the purpose of the check valve in the make-up water line?

5. Did the boiler you worked on have a sight glass for detecting the water level? If so, what was the condition of the inside of the sight glass, dirty or clean?

6. Why is it important for sight glasses to have valves on either side of the glass?